새로 사는 동대문

문구완구시장 속 복합도시

존 홍 서울대학교 교수

NEW FORMS OF LIVING

Mixed-Use Urbanism for Seoul's Historical Dongdaemun Toy and Stationary Market

John Hong Professor, Seoul National University

문구
3676~0801~2 F. 3676~0803
문구 도.소매
삼 영

기호문구
삼영사
742-8989
763-7766
D/C타운
744-4421~2
아트피아
예지사
T.766.3926

완구 할인매장
26-1

5000원

CRAFTBASE
CRAFT
BEER
2018. 10. 24. 수
시대여관

사
745-9857

동해
현대

슈퍼
좌절금지
세종모텔

지봉로
Jibong-ro
9
brother
미싱

DOOSAN
청평화의류도매
동평화패션타운

태광
카렌다사
태광
도장재료
유료
주차장
30분 2000원
인쇄
오가
만물

키스 노래방
온누리 에이전시 직업소개소
945-7255
743-5533
온누리 직업소개소
에이전시 T.945-7255, 743-5533
스포츠 마사지
경락·발관리
crack jack
클랙잭
우림 인쇄
전단지
스티커
각종인쇄
우림
T.742-5518
명 장
키스 노래방
7 ELEVEN
표구사
등산복 등산용품
화랑
칠성유통
각종인쇄 청첩장 봉투 NCR 거래명세서 영수증 판촉물
우림문화사 T.742-5518
칼라 명함 스티커 전단지
스티커
고무인
각종인쇄
거인당
대한민국 인장공예
표구사

83

연구 의제

종말과 잠재력 사이

동대문 문구완구시장

2017년 서울비엔날레 국제 스튜디오의 큐레이터를 맡게 되어 대상지를 선정했을 때, 전제 조건은 서울의 대안적인 도시 재생 가능성을 정의하는 것이었다. 많은 도시 이론가들이 주목해온 것처럼 서울의 오래된 지역은 과거와 현대의 조각보 같다. 서울을 피규어-그라운드 드로잉(Figure-ground drawing)으로 이분해 보면 스케일 차이가 극명하게 드러나기 때문에 한강의 북쪽 지역이 '형편없이' 파편화된 도시계획으로 여겨진다. 서구의 지도 제작 방식을 통해 본 식민지 시대 이전의 골목과 포스트 모던 주거 구획 사이의 균열은 갈등의 서사로 심화한다.

한때 1km 남짓의 선으로 서울을 정확하게 재단했지만, 이제는 도시 조직에 둘러싸이게 된 세운상가 건물을 보면 알 수 있는 것처럼, 도시가 나이 들어감에 따라 이질적으로

보였던 도시계획들이 서로 얽혀가고 있다. 우리가 도시계획가로 일하는 사람이라면 더 정교하고 섬세한 방향으로 서울을 이해할 방법을 개발해야 한다. 한때 도시의 생산과 일상생활의 중심이었던 부분이 노화하면서 이제는 위협을 받고 있다. 생활과 일, 생산과 여가가 서로 다른 영역으로 나눠지고 분산되면서 비효율, 창조성 고갈, 지역사회의 노동력 이탈을 초래했다. 이 연구는 동대문 주변의 여러 지역들처럼 활용도가 낮은 서울 내 경공업 지역들을 조사함으로써 현존하는 물리적, 문화적, 역사적인 자산을 새로운 기술 발전과 결합하여 도시의 기존 영역을 활성화시키는 방안을 제시한다. 서울의 중요한 도시조직 일부를 지우고 다시 설계하는 방법보다는 기존 조직에 복합적인 프로그램을 활용함으로써 보다 의미 있고 생산적인 도시계획을 만들어낼 수 있다.

동대문 문구완구시장 같은 지역들은 역사적으로 수많은 문화 공간과 유산을 보유한 생산의 중심지였지만 지금은 경제적 쇠퇴로 고통받고 있다. 이런 장소들을 걸을 때, 누군가는 십여 년간의 경기 침체기를 겪은 것 같은 녹슨 모습을 마주하게 된다. 어쩌면 우연히 만난 장소일 뿐일지도 모르는 곳이 마치 방금 드나들었던 지역처럼 느껴진다. '비껴간 근대화'는 서울 비엔날레 국제 스튜디오 주요 참여자 마크 브로사(서울시립대학교 교수)가 여러 차례 있었던 라운드 테이블에서 동대문 문구완구시장을 묘사할 때 언급했던 용어이다.

자하 하디드가 설계한 동대문 디자인 플라자와 청계천 복원 사업, 수많은 역사적-현대적 시장과의 인접성에도 동대문 문구완구시장은 이상하게도 '그때 그 시절' 그대로 냉동되어있다. 실제로 이곳에는 여전히 수백 개의 '쪽방(주로 노년층 1인 가구가 사용하는 하위 주거시설)'들이 남아있을 뿐만 아니라, 20년 동안 간판 한 번 바꾼 적 없을 만큼 왕래가 적은 상점들이 자리잡고 있다. 물론, 비주류 힙스터 바와 트렌디한 의류디자이너의 작업실도 있지만, 그 수가 매우 적고 성기게 분포하기 때문인지 탄탄한 지원사업 네트워크라기보다는 경제적 취약층의 그물망에 갇혀있는 것처럼 보인다.

한편, 생산 구역으로서 동대문 문구완구상가는 잠재적인 진원지에 자리 잡고 있다. 창신동에 있는 수천 개의 봉제 공장의 바로 아래이자 세운 전자상가의 동쪽이며 서울의 가장 중요한 가구-골동품 상가 지역 중 하나인 황학동의 서쪽에 위치한다. 엄청난 잠재력과 총체적인 종말 사이에서 창신동 일대는 수십 년 동안 이 애매한 상태에 아슬아슬하게 놓여있다.

이러한 불확실함에서 비롯된 이 책은 동시에 진행된 두 개의 연구를 통해 동대문 문구완구상가를 해석해보려 한다. 첫 번째 연구는 서울대학교의 재능있는 4학년 학생들이 도시 조직의 기존 프로그램을 이용하는 패턴과 방식을 분석하는 과정을 통해 제작한 다이어그램들을 바탕으로 한다. 동대문 문구완구시장에 존재하는 많은 자산을 활용하여 이곳이 창의적인 생산을 위한 새로운 중심지가 될 수 있도록 설정한 사회·문화적 공간을 도시 단면의 형태로 보여준다.

동시에, 큐레이터 임동우 교수(홍익대학교)와 니나 래파포트(예일대학교)의 도시생산주거: Factory for Urban Living 전시에서 대상지가 차세대 인공지능 기술 창출의 중심지가 된다는 사실 기반의 허구를 통해 생산 센터로서 임박한 동대문 문구완구상가의 잠재성을 관찰하는 작업을 그래픽 아티스트 김혜인과 함께했다. '스마트 스피커'와 '스마트 홈'이 더욱더 일상적인 것으로 여겨지면서, 동대문 문구완구상가는 세운 전자 상가와 창신동의 의류 제조업 사이에 균형을 이뤄 새로운 인공지능 장난감을 생산하는 아날로그와 디지털 영역을 거의 완벽하게 결합한다. 본 연구는 동대문 문구완구상가에 잠재된 대규모 사용자 정의기술의 가능성과 도시 한가운데 자리한 도시의 초소형 공장들의 제조업이 곧 맞게 될 미래를 낙관적으로 바라보았지만, 현지 생산이 모든 디지털 정보 네트워크와 전 세계적으로 연결될 때 필연적으로 마주하게 될 프라이버시 손실이라는 암울한 측면 또한 파헤쳤다.

LEGO LEGO LEGO LEGO
승진완구
월드코리아
일방통행

Research Agenda

Between Demise and Potential

The Dongdaemun Toy and Stationery Market

When I served as the curator for the Seoul Biennale International Studios in 2017, the underlying premise for selecting sites within Seoul was their potential to define an alternative kind of urban regeneration. As many urban theorists have noted, Seoul's older regions are a patchwork of ancient and modern. When viewing a binary figure-ground drawing of the areas north of the Han River, the intense shifts in scale amount to what can be read as 'bad' fragmented urbanism. Through the lens of this Western mapping construct, the rift between the pre-colonial alleyways and post-modern housing blocks intensify into a narrative of conflicting opposites.

But as the city ages, these seemingly disparate urbanisms intertwine, as seen in the 1km Seun Arcade building that once cut a clean line through the city but is now encrusted within it. If we are to work as urbanists within this layered history, we need to develop more sophisticated, fine-grain ways of understanding Seoul. Today, these complex zones, once centers of production and daily life, are now threatened. The dispersal of living, work, production, and leisure, zoned into different and defined areas has created inefficiencies, depleted creativity, and taken work away from a local labor force. By examining underutilized light manufacturing spaces in Seoul such as the diverse areas around Dongdaemun, this research reveals how existing physical, cultural, and historical assets can be combined with new technological advances to revitalize existing areas of the city. By leveraging mixed-use programming without resorting to erasing vital parts of Seoul's fabric, more significant and productive regeneration can occur.

Areas such as the Dongdaemun Toy and Stationery Market (DDMSTM) have served as centers of production with a multitude of spatial and cultural assets, but are now suffering from economic decline. When walking through them, one can feel the patina of decades of economic depression, as if the area has been just getting by despite its potentially fortuitous location. Professor Marc Brossa (University of Seoul), one of the important participants of the Seoul Biennale International Studios, brought a succinct term to our roundtable discussions in describing the DDMSTM: the concept of 'Missed Modernization.'

Despite its proximity to the Zaha Hadid designed Dongdaemun Design Plaza, the Cheonggyecheon stream restoration, and numerous historical and modern markets, the DDMSTM remains mysteriously frozen in time. In fact, it is still home to hundreds of one-room 'jjok-bang' (sub-standard residences mostly used by the elderly), as well as low-trafficked shops with storefronts that have not changed in 20 years. There are of course fringes of hipster bars and trendy clothing designers, but these are few and far between and seem to be trapped within a fragile economic web rather than a more robust supporting network.

Meanwhile as a *production* zone, the DDMSTM sits at a potential epicenter: It is just downhill from the thousand or so micro garment factories in Changsindong, it lies just east of the electronics epicenter of the Seun arcade, and is just west of Hwanak-dong, one of the most important antique and furniture markets in Seoul. Poised between incredible potential and total demise, this central site has hovered in this vague state for decades.

It is from this unsure footing that this book examines the DDMSTM through the lens of two concurrent studies: First, through the work of talented 4th-year students at Seoul National University, we analyzed and diagrammed the existing use patterns and programmatic mix. The analysis yielded new architectural sections through the city that demonstrate how the many existing assets can be leveraged to create new spatial and socio-cultural scenarios for creative production.

Concurrently, under curators professor Dongwoo Yim (Hongik University) and Nina Rappaport (Yale University) for the exhibit 'Factory for Urban Living,' I collaborated with graphic artist Hyein Kim to look at the imminent potential of the DDMSTM as a production center through a fact-based-fiction where the area gives rise to the next generation of AI technology. As 'smart speakers' and 'smart homes' are becoming ever more prevalent, the DDMSTM is *almost* perfectly poised between the electronics of the Seun area and the garment manufacturing of Changsindong to produce new AI toys that combine the digital and analog realms. The research took an optimistic look at both the imminent potential of mass customization and micro-factory urban manufacturing latent in the DDMSTM. On the other hand, it also delved into the dystopic side of lost privacy when local production inevitably interfaces with globally omniscient digital information networks.

과거
버려둔 중심지

현재의 동대문 문구완구시장을 포함한 창신동 지역은 서울 성곽의 바로 바깥에 자리하기 때문에 그 자체가 도시 중심지의 한 부분이자 동시에 그곳에 서비스를 제공하는 지역이기도 하다. 창신동 지역의 정치적인 위상을 정의할 때 이 두 가지 속성은 계속해서 두드러진다. 각기 다른 역사적 무대들을 거치는 동안 계속해서 이 지역은 도시에서 일종의 압력 밸브 같은 역할을 맡아왔다. 도심지에 공급할 상품과 음식을 생산했고, 빈곤층에게는 주거를 공급했으며, 동대문의 입구에서 도시와 상업 간의 접점을 마련해주었다.

창신동의 존재는 고려 왕조 시대인 서기 약 1000년쯤의 사료에서도 확인되지만, 서울이 국가의 수도가 되는 1300년대 후기 조선 왕조 설립 이후에 비로소 지리 정치적으로 더욱 중요한 독립체로서 자리 잡는다. '창신동'이라는 이름은 1396년에 인접한 지역인 창인방과 숭신방을 '창신숭인'으로 합병하면서 공식적으로 확립된다.[1] 남쪽으로는 그림 같은 복지천이 흐르고, 북쪽으로는 낙타산(낙타의 등을 의미한다)이 자리한 창신동의 첫 번째 역할은 왕족의 별장을 위해 치외법권인 편안한 장소를 제공하는 것이었다. 맑은 개울과 화강암 산으로 둘러싸인 중심으로 무속적인 힘이 흘러드는 이곳은 시간이 지난 뒤 모순적이게도 자원 추출의 원천이 된다.

17세기 중반, 이 땅에 계속해서 덮쳐온 자연재해로 인해 국가는 기근의 위기를 맞게 된다. 농촌에서 온 실직한 농부들은 국왕이 마련한 긴급 대책을 통해 창신동에 정착할 수 있었다. 이때부터 이 지역의 역할은 휴양지에서 도시의 문젯거리들을 처리하는 곳으로 변화한다. 조선 왕조 초기에 한양 도성 내 인구는 1789년 약 112,000명으로 안정되었던 반면, 창신 지역의 인구는 급속도로 치솟아 76,000명에 육박했다.[2] 번잡해진 종로 거리가 지역의 중심을 가로지르게 되면서 도심지를 보조해야 한다는 과제에 당면하게 되었고 이는 앞서 이루어진 급성장이 가져온 문제들을 덮어두게 했다.

19세기 한양의 모습을 그린 김정호의 〈수선전도〉와 2017년 서울 지도에 드러난 도로망 비교

Comparison of road network revealed in Kim Jung-ho's 〈General Map of Seoul-1860s〉and the map of Seoul in 2017

1899년, 고종이 서대문부터 동대문을 거쳐 청량리에 이르는 철도를 종로를 따라 설치하면서 상업 활동에 박차를 가했다. 철로가 놓인 자리에 있던 창신 지역의 시장이 철거됨과 동시에 이 새로운 도시 기반을 따라 더 큰 규모의 상업이 빠르게 성장했고 자연히 인구 또한 증가했다. 이로써 배오개 시장은 현재 동대문 시장의 시초가 되었다. 이곳에서 채소와 같은 농산물을 포함한 많은 상품이 거래되었고 이는 여전히 한양 도성 내부의 인구를 지탱하는 것이었다.

1900년대 초기에 이르러 예지동(현재의 세운상가 지역)의 광장 주식회사는 천과 리넨 관련 제조업의 중심이 되었다. 그 결과 창신동은 직물 산업 종사 노동자 계층이 집중된 지역이 되었고 이 때문에 근대사에서 의류 산업 지역이라고 기록된다. 화학 및 기계를 포함하는 다른 공장들은 왕산로라고 하는 주요 거리를 따라 섞여서 분포했다. 일제에 합병된 이후, 창신동 일대가 조선에서 세 번째로 인구 밀도가 높은 지역이 됨에 따라 지역의 빠른 고밀화를 수용하기 위해 식민 정부는 여러 번에 걸쳐 필지를 쪼갰다.[3] 제2차 세계 대전 동안 벌어진 일제의 집약적인 활동으로 인해 창신동의 자원과 노동력은 극단적으로 착취되었고 그에 따라 서울 또한 빈곤에 처하게 된다.

1944년에는 88,000호가 넘는 주거 단위가 부족해졌고, 총 40%에 이르는 가정이 집 없이 살았다. 이에 대한 대책으로 일제 정부는 필지를 더 작게 나누게 되었고 그리하여 오늘날 우리가 만나는 고밀한 도시조직이 이루어졌다.[4] 1950년 한국 전쟁이 또 다른 고난의 물결을 몰고 오면서 지방에서 심지어 더 많은 인구가 떠밀려와 수많은 불법 주택 정착지를 형성한다. 심각해진 상황을 통제하기 위한 노력의 일환으로 서울 최초의 중복도식 아파트인 '동대문 아파트'가 건설되었고 지금도 동대문 문구완구시장의 다양한 건물군 중 하나로 남아있다.[5] 그러나 위기를 무마하기 위한 몇 가지 시도들이 있었음에도 전후 시기는 불법 주택, 도시 빈곤층에게 닥친 역경, 부동산 투기 그리고 무분별한 철거와의 계속된 투쟁으로 기억된다.

이렇게 많은 역경을 겪어왔음에도 감당할만한 가격대의 의복에 대한 기본적인 수요는 계속 존재해왔기에 전쟁 이후에 이르면 의류 시장은 점차 성장세를 보인다. 1955년 국내에서 판매된 의류 총량의 무려 60%가 창신동에서 제조되었다. 1961년에는 창신동 지역의 가장 남쪽에 평화 시장이라는 선형의 거대 고층 건물이 건설되었고 이 선례를 따라 다른 많은 시장이 세워졌다. 그러나 창신동은 1977년에 이르러 그 성장의 한계에 다다랐다. 이 지역에서 생산되는 상품은 매우 적은 이익을 남기는 일상용품이었기 때문에, 노동자를 위한 기초적인 수준의 주택을 공급하는 데에 어려움이 따랐다. 앞서 말한 성공의 이면에서는 수많은 젊은 여공들이 저임금을 받으면서 봉제 공장 상층부에 있는 다락방에서 열악한 생활을 하고 있었다.

1960년대, 일본인들이 자주 다니던 집창촌의 영업이 금지되어 가는 한편 미군이 차지하고 있던 여관 대부분이 빛이 들지 않는 좁고 열악한 쪽방들로 바뀌어 현재까지도 그 모습이 그대로 남아있다.[6] 실제로 동대문 문구완구시장은 쪽방처럼 질 낮은 주거 형태들이 가장 밀집되어있는 서울의 5개 지역 중의 하나이다. 1970년대에는 계속해서 서서히 쇠퇴해가는 창신동이 목격된다. 도시 개발의 초점이 한강 이남에서 새롭게 떠오른 강남권으로 옮겨지면서 더 나은 교육과 주거에 접근할 기회가 남쪽으로 이전되었고, 이로 인해 동대문 문구완구시장은 '근대화가 비껴간' 지역이 되었다. 작은 사업체들로 이루어진 다양한 사회 기반 시설들을 무시한 채로 대형 의류 제조업자들은 시장의 공공재를 장악하기 시작했다. 작은 사업체들은 그 과정에서 노동조합 분쟁에서 비롯된 문제들을 겪으며 더욱 파편화되었다.[7] 1990년대 중반에는 중국의 저렴한 노동력이 유입되면서 'IMF 위기'로 알려진 경기 침체가 상황을 더욱 악화시켰다.

오늘날에도 의류 산업은 여전히 물 새듯 천천히 쇠퇴해가고 있는데, 이는 특히 전체 경제 성장률이 감소한 것과 연관이 깊다. 물론 국내 의류 판매를 위한 기반 시설을 제공하는 두산 타워나 청계천 복원 사업과 같은 도시 재생 등, 긍정적인 자금 투입도 있어서 그 쇠퇴는 어느 정도 상쇄되어왔다. 게다가 2007년에 이 지역은 작은 규모의 개입들을 통해 도시를 살리는 침술 요법을 촉발하는 역할을 맡아 서울의 재활성화 지역으로 승격되었다. 그렇기는 하지만 다소 간편하게 지어진 '브랜드 네임'(완구와 문구 사이의 관계성은 크게 깊지 않아 보인다)이 씌워지자 지역의 정체성이 결여되었고 그 결과 사회의 주목을 덜 받게 되었다. 지금도 창신동에는 역사적인 관성이 남아있는데 이는 창신동의 가장 강력한 잠재력이면서도 자신의 가장 지독한 적수이다.

Past

A Discarded Center

The political status of the Changsindong region, including the area that is now the Dongdaemun Toy and Stationery Market (DDMSTM), as just outside of the historical city wall has defined its constantly emerging condition: It is both directly part to the center and servile to it. Through different historical stages and manifestations, it has acted as a kind of urban pressure valve: producing goods and food for the city core, housing its poor, and supplying a commercial interface at the mouth of the historical Dongdaemun (east gate).

Changsindong's existence can be traced back to approximately 1000 AD during the Goryeo Dynasty, but it was the establishment of the Joseon Dynasty in the late 1300s that allowed it to become a more prominent geopolitical entity when Seoul became the nation's capital. The name 'Changsindong,' formally established in 1396, combined the names of adjacent Changinbang and Soongsinbang, amalgamating them into 'Chagsinsoongin.'[1] With the picturesque Bokchacheon stream to its south and the Naktasan (meaning camel back) mountain to its north, its first extra-territorial role was as a convenient place for the vacation houses of the royal families. Its shamanistic energy flows centered around clean waterways and the granite mountain which paradoxically became the sources of resource extraction later on.

In the mid 17thc, natural disasters continually put Korea in crises of famine. Through emergency policies set up by the king, out-of-work farmers from the countryside were allowed to settle in Changsindong. This was the beginning of the region's shift from a place of leisure to a place that absorbed urban woes. Even as the population inside Seoul's historical wall during the beginning of the Joseon Dynasty remained stable at approximately 112,000 people in 1789, in comparison, the population in Changsin rapidly grew to 76,000.[2] With the busy east-west Chongnro street cutting through its center, its immediate relationship as serving the city core harbored this growth.

Further prompting commercial activity was the construction of a rail built by King Gohjong in 1899 connecting Soedaemun (the west gate) via Jongrno through Dongdaemun out to Cheongrangyi. Even as markets in Changsin were first demolished along the railway right of way, larger-scale commercial businesses quickly grew around this new infrastructure along with a growth in population. In this regard, Baeohgae market became the seed of the current Dongdaemun market - many items including vegetables were sold there, again in a servile role to the population inside the city wall.

By the early 1900s, Gwangjang corporation in Yejidong (currently the Seun-sangga area) became the center of cotton and linen manufacturing. As a result, Changsindong became a concentrated working class zone for cotton production marking its modern history as a garment production region. Other factories including chemical and machinery were intermixed along the main street, then called Wangsan-ro street. As the population expanded into becoming the 3rd largest area of Koreans following Japan's annexation of Korea, properties were continually subdivided by the colonialist government to accommodate the rapid densification.[3] Japan's intensive activities during World War II forced Seoul into impoverishment as Changsindong becoming a place of extreme resource and labor extraction.

By 1944, there was a lack of over 88,000 dwelling units amounting to 40% of families without housing. In response, the Japanese government subdivided the properties even further resulting in the creation of the dense fabric that we see today.[4] Following another wave of hardship from the Korean war in the 1950, many illegal housing settlements formed as even more people migrated from the countryside. In an effort to curb the negative conditions, the first double-loaded corridor apartment in Seoul called the 'Dongdaemun Apartment' was constructed and still exists among DDMSTM's diverse buildings.[5] Despite these limited attempts to relieve the situation, the post-war period was marked by a constant struggle between illegal housing, the plight of the urban poor, real estate speculation, and unchecked demolition.

Even with these hardships, the basic need for affordable clothing allowed the garment markets to expand after the war. In 1955, 60% of apparel sold domestically was made in Changsindong. At the south end of the site, the linear megastructure of Pyeonghwa Market was constructed in 1961 with the rise of many other markets following suit. But by 1977 the area had reached its econimic height. Because the support items were made within slim profits to everyday use, the underside of this success was the constant struggle for egalitarian housing, a need that continues today. Many young women factory laborers working at low wages occupied the sub-standard attic spaces above sewing factories.

Meanwhile the banning of prostitution areas frequented by Japanese, and subsequently American soldiers in the 1960s resulted in the conversion of the Yeogwan they occupied (small temporary housing) into narrow, low-standard, many times light-less Jjok-bang (one-room housing) that still exist today.[6] In fact, the DDMSTM is one of the five areas in Seoul with a high concentration of this degraded housing type.

The 1970s witnessed the further slow decline of Changsindong. As urban development focused on the new Gangnam areas south of the river, the DDMSTM site became a place of 'missed modernization' as better educational and housing opportunities relocated south. Bypassing the diverse infrastructure of small companies, larger clothes manufacturers began to seize the market shares, while smaller businesses became more fragmented by complications stemming from labor union disputes.[7] In the mid-1990s, the economic downturn known as the 'IMF crises' in conjunction with cheap labor from China would further exacerbate the situation.

Today the steady trickle of this garment industry's recession is still evident, especially as the economy as a whole slows. Positive injections such as the Doosan tower which provides a domestic selling infrastructure have offset the decline, as well as urban regeneration including the award-winning restoration of the Cheonggycheon stream at the site's southern border. Moreover, in 2007, the area was promoted as a revitalization district sparking the urban acupuncture of small-scale interventions. Nonetheless, a lack of identity for the region overlaid with an imposed simplified 'branding' (the relationship between toys and stationery seem to lack significance) have brought limited public notice. The site's historical momentum remains its greatest potential and its own worst enemy…

미주
Footnotes

1 강홍빈, 『창신동-공간과 일상』, 2011, 서울역사박물관, p.22.
2 앞의 책, pp.23.
3 손정목, 『일제강점기 도시화과정연구』, 1996, 일지사, p.145.
4 서울육백년사 홈페이지 (시대사〉일제침략하의 서울〉시민생활〉주택사정〉토막과 토막민)
5 앞의 책, pp.53.
6 동아일보, "昌信洞(창신동)에 큰 불", 1965.12.17, 제3면.
7 앞의 책, pp.65.

지역경계선
Regional boundary

학교
School

제단
Shrine

경전철
Light rail transit

철도
Railway

참고: 서울특별시가지도 (1958)
Reference: Map of Seoul (1958)

채석장
Quarry
서울성곽
Seoul Fortress
흥인지문
Heung-injimun
복자천
Bokja-cheon(stream)
청계천
Cheonggye-cheon(stream)

현재
숨어있는 가능성

동대문 문구완구시장을 거닐며 마주하는 모든 모퉁이마다 역설이 발생한다. 이 지역은 역사적인 동대문 시장과 희미하게 빛나는 동대문 디자인 플라자에 인접해 있지만 수십 년간 경기 침체에 빠진 것으로 보인다.〈그림1〉 다양한 한옥과 역사적 구조물이 모여있지만, 대부분이 반(半) 불법인 '쪽방'으로, 빛과 공기, 위생에 대한 기본적인 접근이 결여된 아주 작은 공간들이다.〈그림 3〉 문구완구시장의 두 가지 소매 품목 모두 초등학교에 다니는 어린이들이 주요 소비 주체지만, 이 지역 인구의 평균 연령은 서울에서 가장 높은 편에 속한다. 또한, 여기는 1인 가구가 가장 많이 모인 지역 중 한 곳이다〈그림 4〉: 어린이들이 거주지에서 공립 학교, 학원(또는 방과 후 교실)을 오고 가며 가로지르는 지역들의 중심에 위치함에도 여기에 사는 가족은 거의 없다.

동대문 문구완구시장의 '시대에 뒤쳐진' 사업 활동이 느린 쇠퇴 상태에 머무르고 있다는 점은 '유리잔에 물이 반이나 들어있다'는 생각의 증거이다. 이 지역의 긍정적인 잠재력은 일상생활의 기초가 된다. 일례로 소매업과 다른 사업들이 전반적으로 쇠퇴하는 동안에도 한국의 의류 산업은 꾸준히 성장하고 있다. 이론상 동대문 문구완구시장이 속한 창신동 지역 전체가 다수의 소규모 봉제산업으로 인해 이익을 얻을 수 있게된다. 우리가 반쯤 찬 유리잔을 분석의 출발점으로 삼는다면, 현재 사이트에 숨어있는 가능성을 확인할 수 있을 것이다. 아직 모두 종합된 것은 아니지만, 동대문 문구완구시장의 경제적-문화적 잠재력을 활용할 강력한 지표로 작용할 수 있는 오늘날의 주요 요인 다섯 가지는 다음과 같다:

1. 높은 교통 접근성, 편리한 교통수단 이용
2. 고령화 인구
3. 의류 제조 산업의 회복력
4. 1인 가구의 증가
5. 새로운 형태의 통합된 기술을 일상생활에 도입하려는 시도들에 대한 현대 소비자들의 저항

그림 1. 서울시 소매업 및 도매업의 폐업률 추이

Figure 1. Seoul retail and wholesaler closure rate trend

통계청, 전국사업체조사, KOSIS (2013)

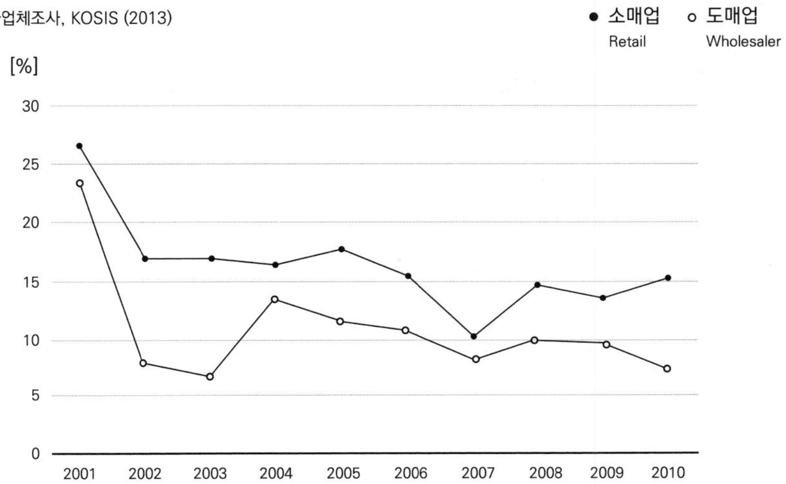

그림 2. 제조업의 산업형태별 비중

Figure 2. The proportion of manufacturing industry by type

국가통계포털 (2014)

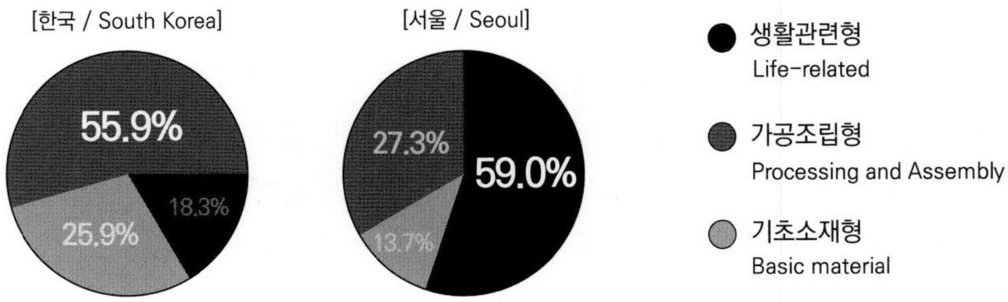

그림 3. 평면도로 본 쪽방

Figure 3. Example: Plans of Jjokbang

AURI 건축도시포럼 #2. 역사성과 시간이 녹아있는 주거지 (2010)

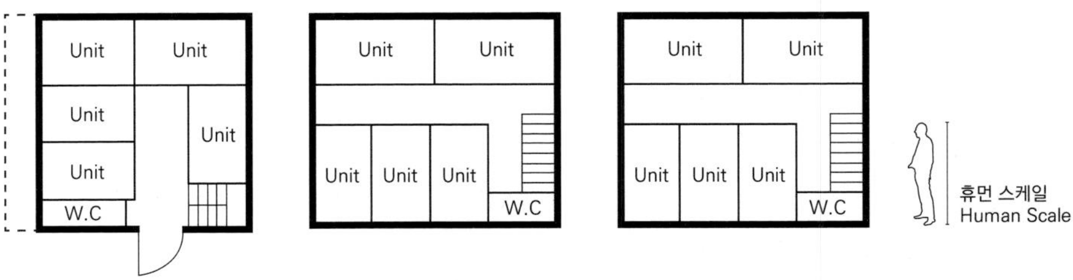

첫 번째 요인은 역사적인 도심 지역과 근접해있다는 점뿐만 아니라 사이트의 북서쪽과 북동쪽 코너의 지하철에 쉽게 접근할 수 있다는 점 또한 포함한다. 1호선과 4호선, 1호선과 6호선이 교차하는 곳으로 두 역 모두 환승 지점이며, 서울의 첫 지하철인 1호선은 동쪽과 서쪽 지역 대부분의 역사적인 장소를 연결한다. 지역의 서쪽을 지나는 4호선은 교통의 중심인 서울역과 남서쪽으로 연결되어있다. 한편, 북서쪽에서 북동쪽으로 이어진 6호선은 홍익대학교나 고려대학교 등 주요 교육 허브를 연결한다.〈지도 1〉 이처럼 다양한 도시 관계들이 모여있기 때문에 문화, 역사, 상업 및 교육 등 모든 개발 요인들 간의 관계를 쉽게 추적해갈 수 있다.

인구 통계자료 및 의류 산업의 탄력성의 측면에서 두 번째와 세 번째 요인을 살펴보면, 의류 및 신발, 인쇄 등을 포함한 이 지역의 다양한 제조기술 분야에 종사하는 고령 근로자를 관리하는 것은 진귀한 '노하우'를 축적하는 것과 같은 의미가 있다. 수공업 기술이 사라져가는 동안, 젊은 세대 안에서는 '아날로그'가 다시 유행하거나 심지어는 페티쉬의 양상으로 늘어나고 있다. 특히 다른 사업들의 쇠퇴에도 의류 판매가 증가하고 있는 최근 데이터를 고려했을 때,〈그림 6〉 동대문 문구완구시장은 수공업 기술이 새로운 세대의 기업가에게 전달되는 장소가 될 수 있다.

네 번째 요인은 1인 가구가 집중되고 서울의 1인 주거 유형에 대한 수요가 점차 증가하는 것에서 기인한다.〈그림 7〉 창신동 일대는 새로운 세대의 1인 가구들을 기존의 노인 공동체와 함께 수용할 수 있다. 이것은 직장 또는 다른 형태의 집단이 내부의 상호작용을 통해 노하우를 전달하는 과정에서 시너지 효과를 낼 수 있다. 새로운 협업 네트워크, 멘토링 기회와 더불어 가족 형태를 띤 유사 집단에도 적용될 수 있다. 이곳은 이미 소규모 도시 조직으로 이루어져 있기 때문에, 더 큰 공간들로 구축된 서울의 다른 지역보다 공간을 세분화해야 하는 문제가 덜하다.

한국이 기술에 빠르게 적응하는 것으로 알려진 것과 달리 일반 대중들은 현재 기술에 만족하고 있거나, 새로운 기술이 가져오는 복잡성에 적응하는 과정에서 점점 더 불안해지는 지점에 이르렀다는 것이 다섯 번째 요인이다.〈그림 10〉 이것을 대량의 신기술을 수용하는 일의 대항세력으로 해석한다면 아날로그의 부활 혹은 새로운 하이브리드화가 일어날 수도 있다. 흥미롭게도 구글과 아마존 같은 기술 거인들은

그림 4. 인구 밀도 비교

Figure 4. Comparisons in population density

세계 대도시 비교 - 서울연구원, 12개 주요 대도시 통계현황 (2008)
World Cities

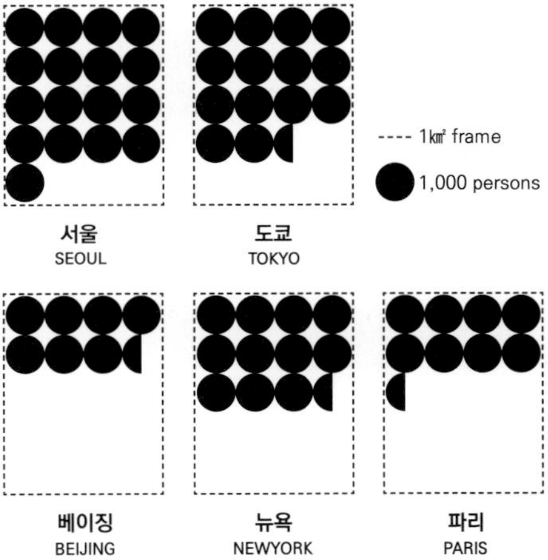

종로구 동별 1인 가구 수- 서울특별시, 주민등록인구 (2010)
Jongno-gu one person households

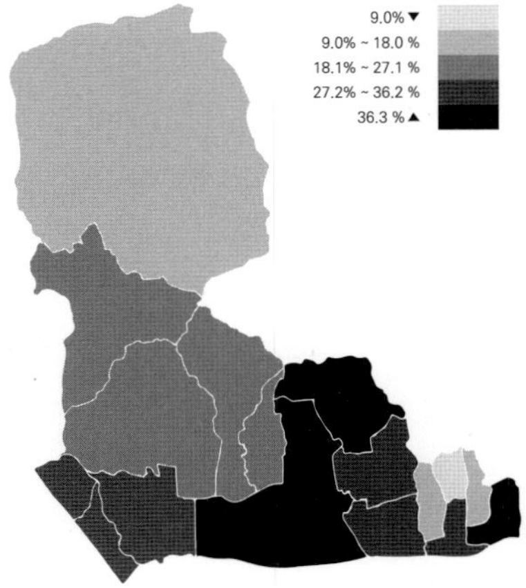

그림 5. 서울시와 창신동의 쪽방 및 거주자 현황

Figure 5. Jjokbang and residents status in Seoul & Changsin-dong

서울특별시, 서울통계, 주민등록인구 (2016)

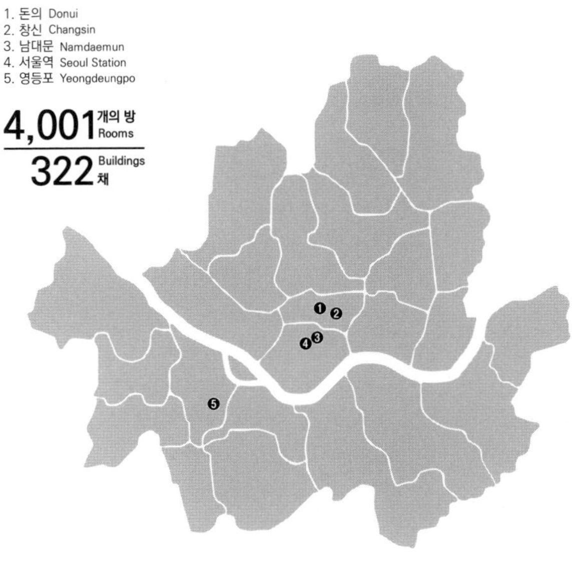

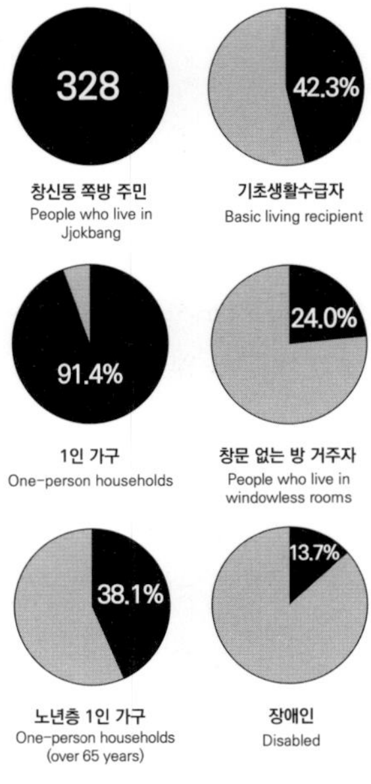

끊임없이 잔물결이 이는 미시 사회의 변화에는 침투하기가 어렵다는 사실을 한국 문화를 통해 알아냈다. 스마트 스피커와 같은 미국의 유비쿼터스 제품들은 한국 시장에 흡수되기 시작했다. 이러한 방식으로, 초소형 제조업은 공예와 기술을 결합한 제품들을 이용해 대기업의 물건들보다 더 빠르게 트렌드에 대응할 수 있다.

만약 이 지역이 실패하게 된다면 그 원인은 무엇일까? 기술적인 원인도 있고, 사회적인 원인도 있다. 거대한 스케일의 도로와 인프라스트럭처가 상품을 사이트의 주변으로 가져올 수 있지만, 블록 내부의 원형 기하 구조 및 작은 스케일의 도로는 물품의 원만한 운반을 어렵게 한다. 오토바이와 보행자가 충돌하는 혼잡한 상황에서 볼 수 있듯 보행자의 통행량에 의존하는 상업과 차량에 의존하는 배송 및 제조업은 병목현상을 발생시켜 잠재력을 제한한다. 그러므로 세밀한 도시 구조를 제거하지 않고 다양한 형태의 배송 연결망을 만들어야 한다. 사회적 경계의 측면에서 보면, 고령 인구는 변화에 저항하는 힘을 만들어 낼 수 있다. 오래된 삶의 방식은 본질적으로 (모든 비효율적인) 도시의 구조와 그 관습에 내재하고 있다. 이 경우, 오래된 것과 새로운 것의 사이를 지우는 것보다는 오히려 덧씌우는 방법들이 필요하다.

그림 6. 서울시 상위 5대 제조업의 구조변화

Figure 6. Changes in Seoul's Top five manufacturing industries structure

통계청, 전국사업체조사, KOSIS, 해당 연도

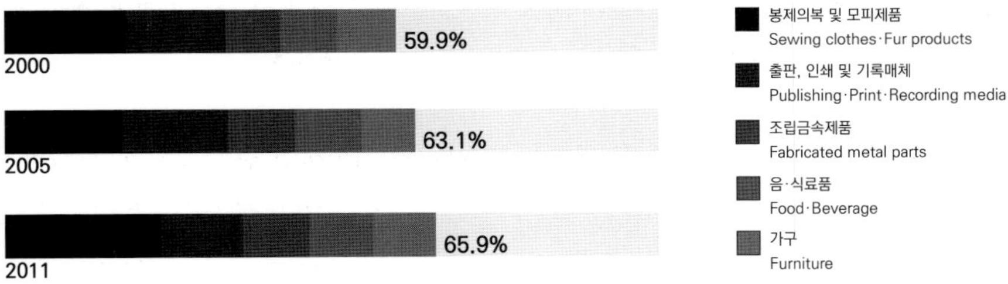

그림 7. 서울시 가구당 평균 가구원 수와 1인 가구 비율

Figure 7. The average number of households and the ratio of one person households in Seoul

통계청 (발표 년도: 2011)

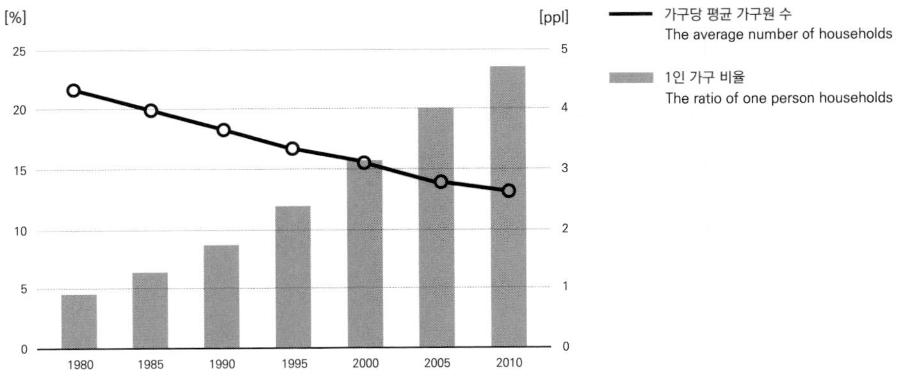

그림 8. 서울시 1인 가구 점유형태 변화

Figure 8. Change in residence pattern of one-person households in Seoul

국토해양부, 주거실태조사, 해당 연도

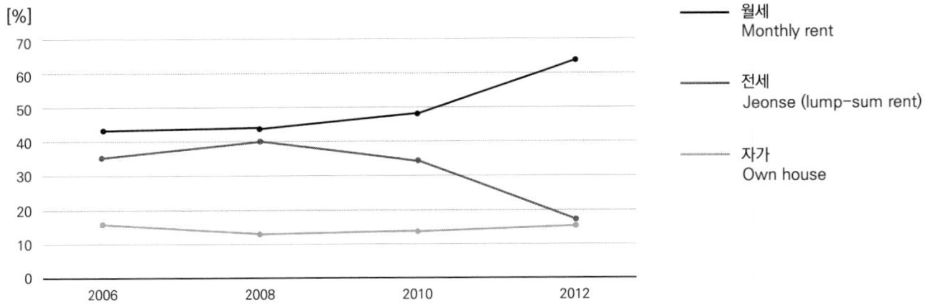

Present

Latent Potential

When walking through the DDMSTM site, at every corner one is confronted with a paradox. The area is adjacent to the historical Dongdaemun market and glimmering Dongdaemun Design Plaza, yet appears to be in a decades-long economic recession.[fig 1] There are a diverse collection of hanok and historical structures, yet most have been subdivided into semi-illegal 'jjok-bang,' tiny rooms that lack the basic access to light, air, and sanitation.[fig 3] The Toy and Stationery market, both retail items that cater to school-age children are two of the main economic drivers, but the average age of the local population is one of the highest in Seoul.[fig 4] It also has one of the highest concentrations of one-person households: while children cross through the site back and forth from residences, public schools, and hagwon (after-school education), few families live here despite its central location.

The fact that the 'obsolete' business activities in the DDMSTM area have remained in a state of slow decline is a testament to the idea that the 'glass is half-full.' The positive potential of the area underlies its day-to-day existence. For instance, even as the number of retailers and businesses in Korea on the whole are decreasing, business associated with the garment industry have consistently been on the rise — theoretically, the whole Changsindong region where the DDMSTM is a part, should be benefitting from this. Thus, if we take the glass half-full as an analytical starting point, we can identify the present latent potential in the site.

Five main present-day factors, although not yet converging, could act as a powerful indicator to how the DDMSTM's economic and cultural potential could be tapped:

1. Access to convenient transportation
2. The aging population
3. The resilience of the garment manufacturing industry
4. The rise in single-person households
5. Consumers' recent resistance toward incorporating new forms of technology into their daily lives.

In the first case, besides its obvious proximity to the historical city core, there is ample subway access in the north-west and north-east corners of the site, both of which are transfer stations that combine subway lines 1, 4, and 6. Line 1 as the first subway in Seoul connects most of the city is historical sites east and west. Line 4 connects westward through to the southwest making an important connection to Seoul Station. Meanwhile line 6 sweeps from the northwest to the northeast connecting major educational

그림 9. 관심 있는 스마트제조기술 Figure 9. What kinds of smart technologies are you interested in?

서울연구원, 「4차 산업혁명 시대, 서울 제조업의 특성과 정책제언」 (2017)

※ 기업체 응답 / 복수응답
Responses from companies,
Allow multiple responses

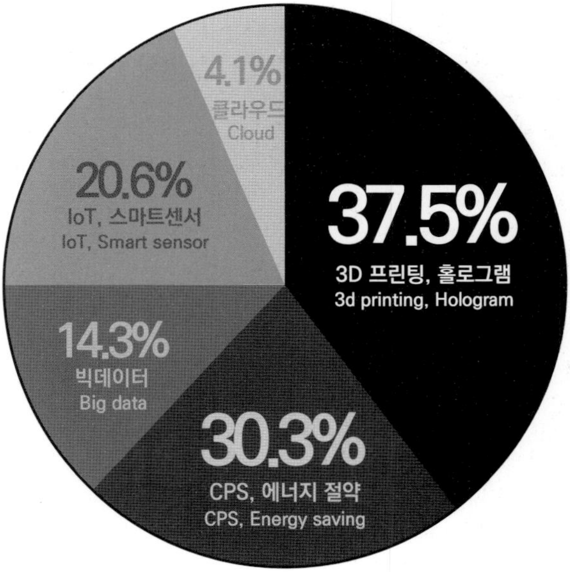

그림 10. 업체의 스마트제조 도입장벽 Figure 10. Why didn't you apply the smart technologies to manufacturing business?

서울연구원, 「4차 산업혁명 시대, 서울 제조업의 특성과 정책제언」 (2017)

※ 단위 (Unit): %

※ 기업체 응답 / 복수응답
Responses from employees, Allow multiple responses

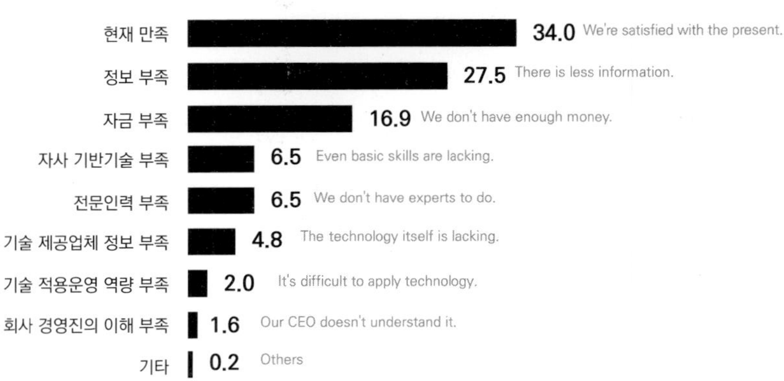

hubs such as Hongik University and Korea University,⟨map 1⟩ Through the collection of these diverse urban relationships, one can easily trace lines between all the city's developmental sectors of culture, history, commerce, and education.

In terms of the second and third point of demographics and the resilience of the garment industry, the various craft manufacturing sectors in the area including garments, shoes, and printing along with the elderly workforce that manages it, equates to a rare accumulation of 'know-how.' While hand-craft knowledge is decreasing, there is a resurgent increase (or even fetish) of the 'analog' among the younger generations. DDMSTM could become a place where the techniques of the hand-made could be passed along to a new collective of entrepreneurs, especially in light of recent data where garment sales are on the rise despite the drop in other businesses.⟨fig 6⟩

Fourthly, because the concentration of one-person households and the rise of the need for this type of housing in Seoul generally,⟨fig 7⟩ the site could house a new generation of single-person households alongside the existing elderly community. This could perform synergistically with the passing on of know-how mentioned above through hybrid living/workplace interactions and other forms of collective community events. New collaborative networks, mentoring opportunities, and even pseudo-family structures could be cast across generational lines. Because of the already small-scale fabric, the issue of subdividing spaces would be less intrusive than in other parts of Seoul that have been built up with larger spaces.

To the fifth point, even as Korea is known for its quick adaption of technology, we have reached a leveling-off point where the general public is either satisfied with the present or increasingly apprehensive with adapting to the complexity that new technology brings.⟨fig 10⟩ If this can be interpreted as a counterbalance against a wholesale acceptance of the new, there could well be a resurgence of the analog, or better yet a new hybridization with it. Interestingly, tech-giants like Google and Amazon find it difficult to penetrate the micro-societal shifts that constantly ripple through Korean culture. Items like smart speakers, ubiquitous in the U.S. have only started to be absorbed into the Korean market. In this way, micro-manufacturers can respond much more quickly to trends than the big brands through customized products that combine craft and tech.

What are the ways the region can fail? Some are technical while others are social. While large-scale roadways and infrastructure can bring goods to the site perimeter, the circuitous geometry and small-scale alleyways within the interior of the block do not allow for a smooth transfer of these goods. One can see in the congested conflict between mopeds and pedestrians, a bottle-neck that limits business transactions that rely on pedestrian foot traffic as well as delivery and manufacturing reliant on vehicles. Therefore, without resorting to demolishing the fine-grain fabric that endows the area with potential, knew forms of delivery networks must be created.

Finally, in terms of social boundaries, the aging population can also create a force of resistance toward change. Old ways of life are inherently embedded within both the physical fabric of the city (with all its inefficiencies) and its social practices. In this case there should be ways of overlay, rather than erasure, between old and new.

Map 1.

교통수단과 녹지 분포

Transportation and Green Areas

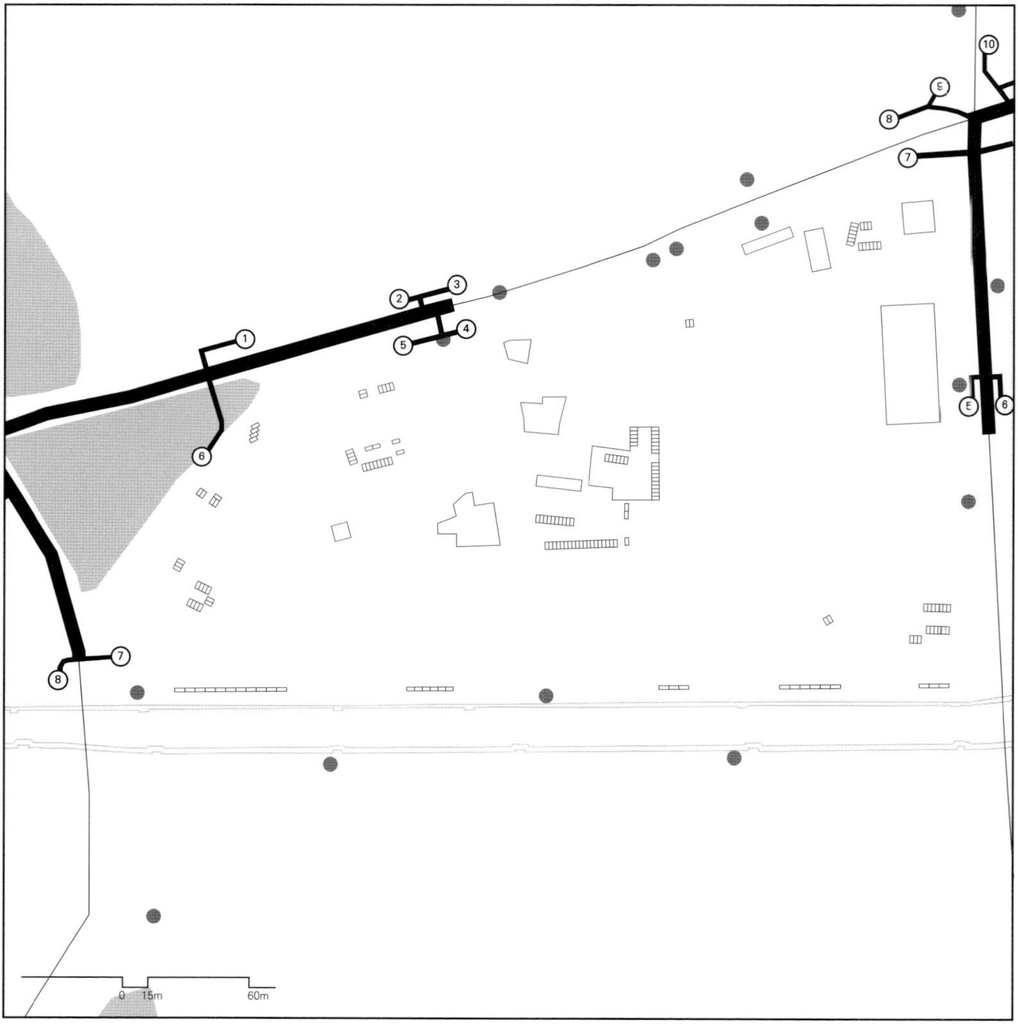

■ 지하철 Subway　　녹지 Green Area　　● 버스정류장 Bus-stop　　□ 주차장 Parking-lot

Map 2.

침투불가능한 영역
Impermeable Areas

침투불가능 Impermeable　　침투가능 Permeable

Map 3.

이용 경로

Path Use

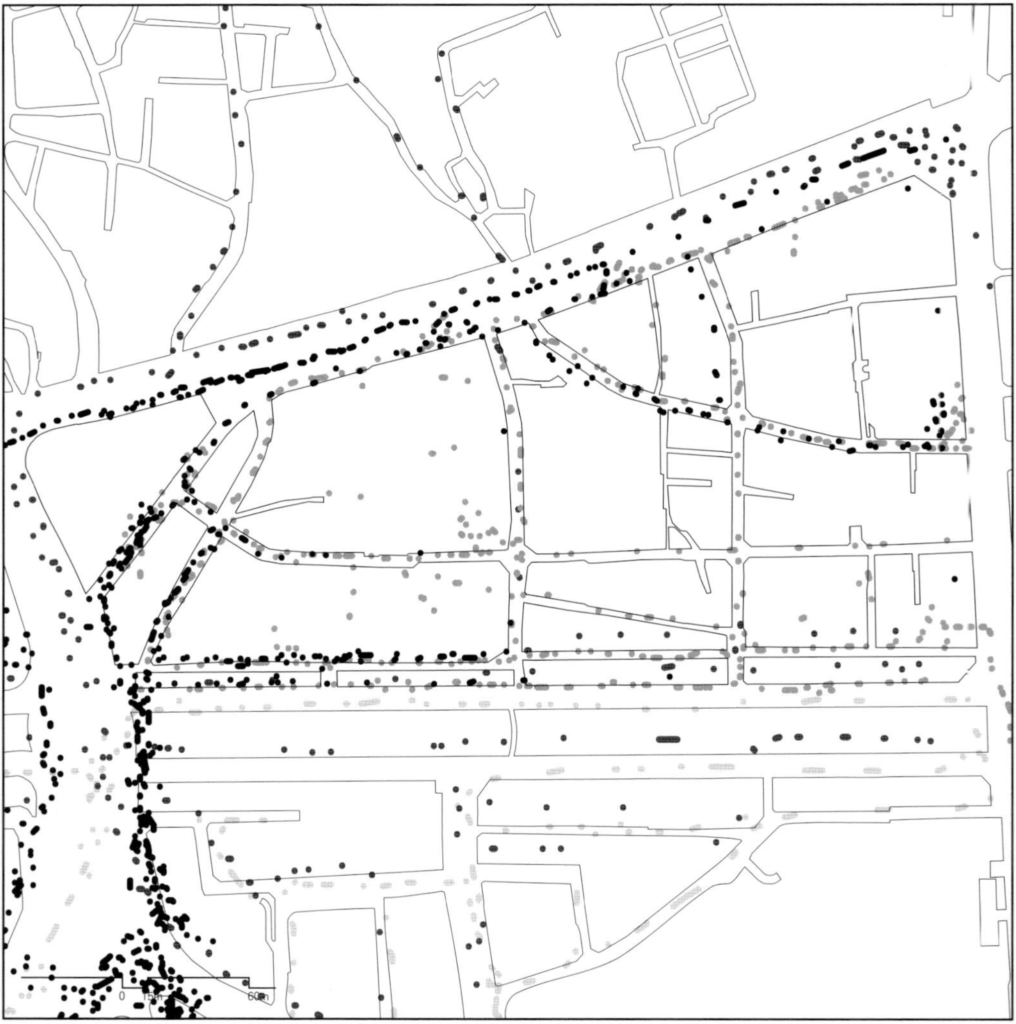

● 방문자, 여행자 Visitor, Tourist
● 의류 제조업 종사자 Clothing Production
● 대상지 내의 상인 Merchant in the Site
● 도매 소비자 Wholesale Consumer

Map 4.

도로 상태

Road Condition

운전 가능 Drivable — 도보 가능 Walkable — 가로등 Street Light — 연결된 블록 Connected Block

Map 5.

운송 수단과 창고

Delivery Vehicle and Warehouse

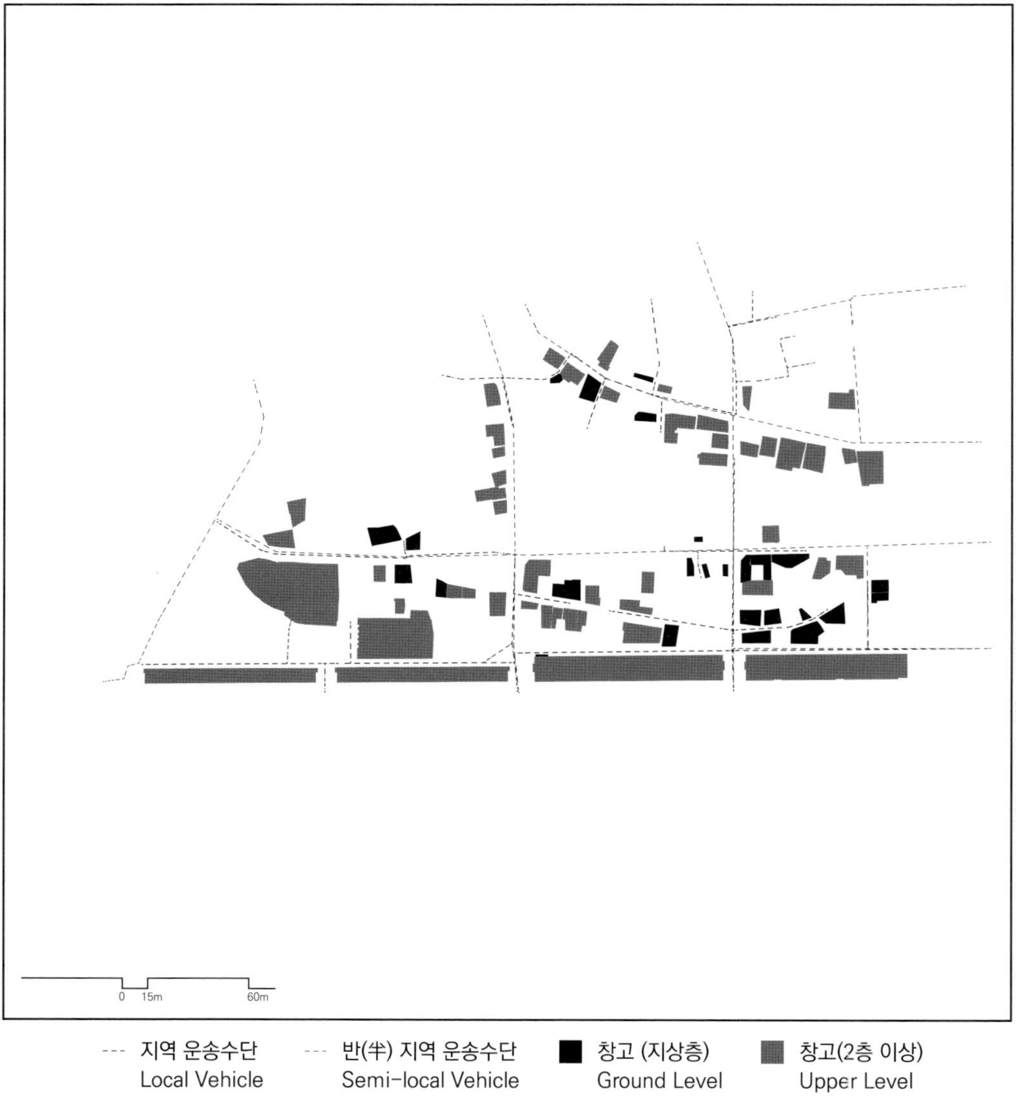

Map 6.

인쇄소, 완구점과 문구점

Print Shop, Toy and Stationary Market

인쇄 Print — 관련 제조업 Manufacturing — 관련 도매업 Wholesale

완구 & 문구 Toy & Stationary: 소매업 Retail, 도매업 Wholesale

Map 7.

제화 산업

Shoes Industries

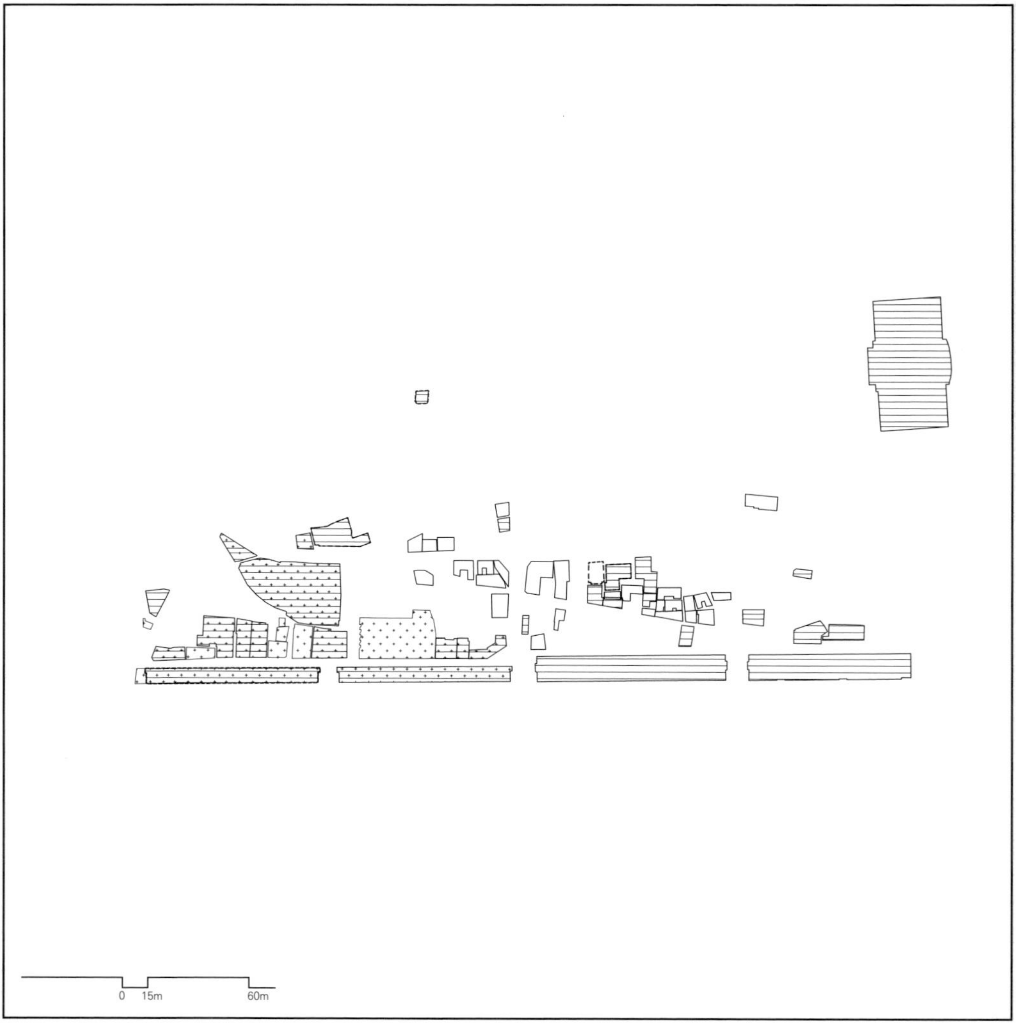

--- 신발 제조업 Shoes Manufacturing
— 자재 제조업 Material Manufacturing
신발 소매업 Shoes Retail

Map 8.

숙박시설과 거주시설

Accomodation and Residence

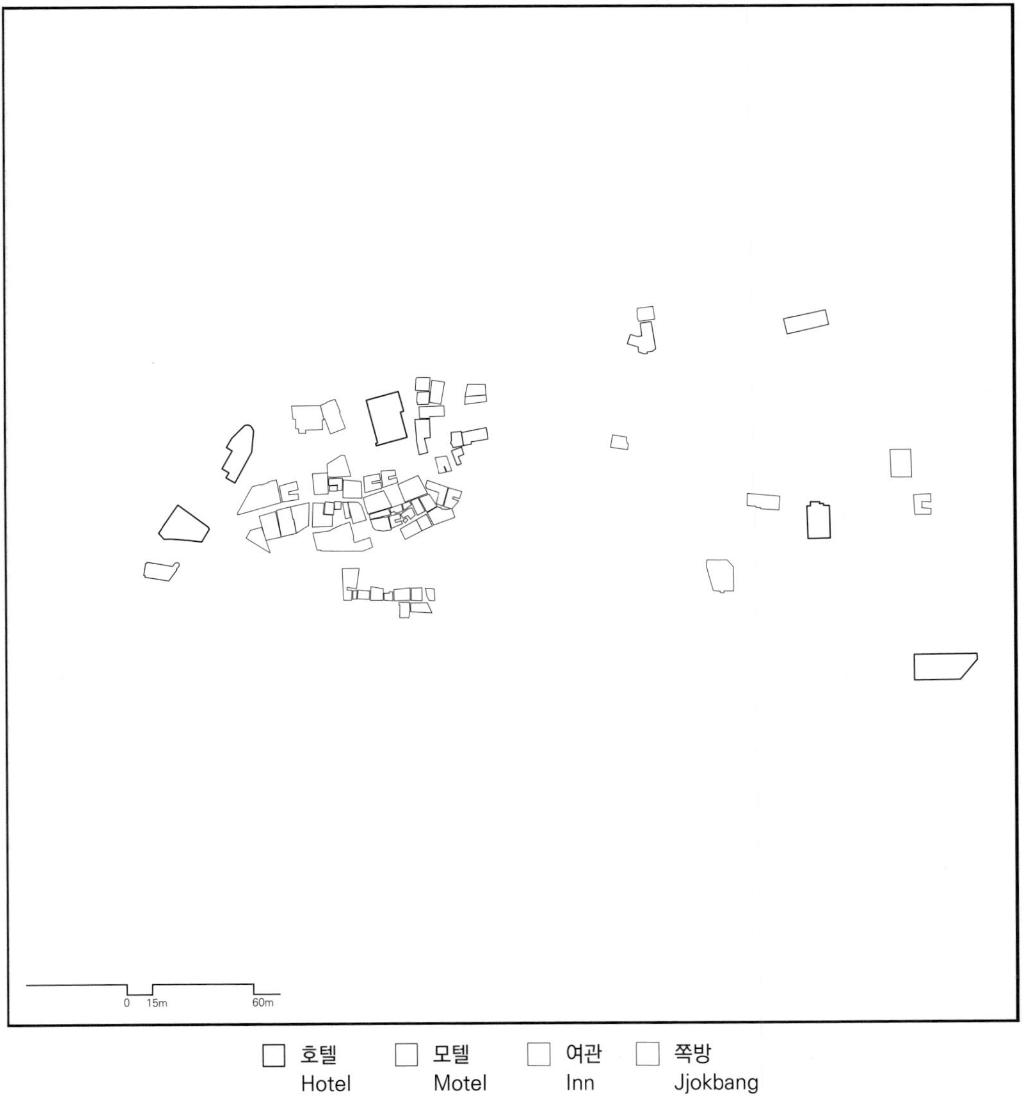

<u>Map 9.</u>

가장자리 건물들의 높이
Edge Buildings Height

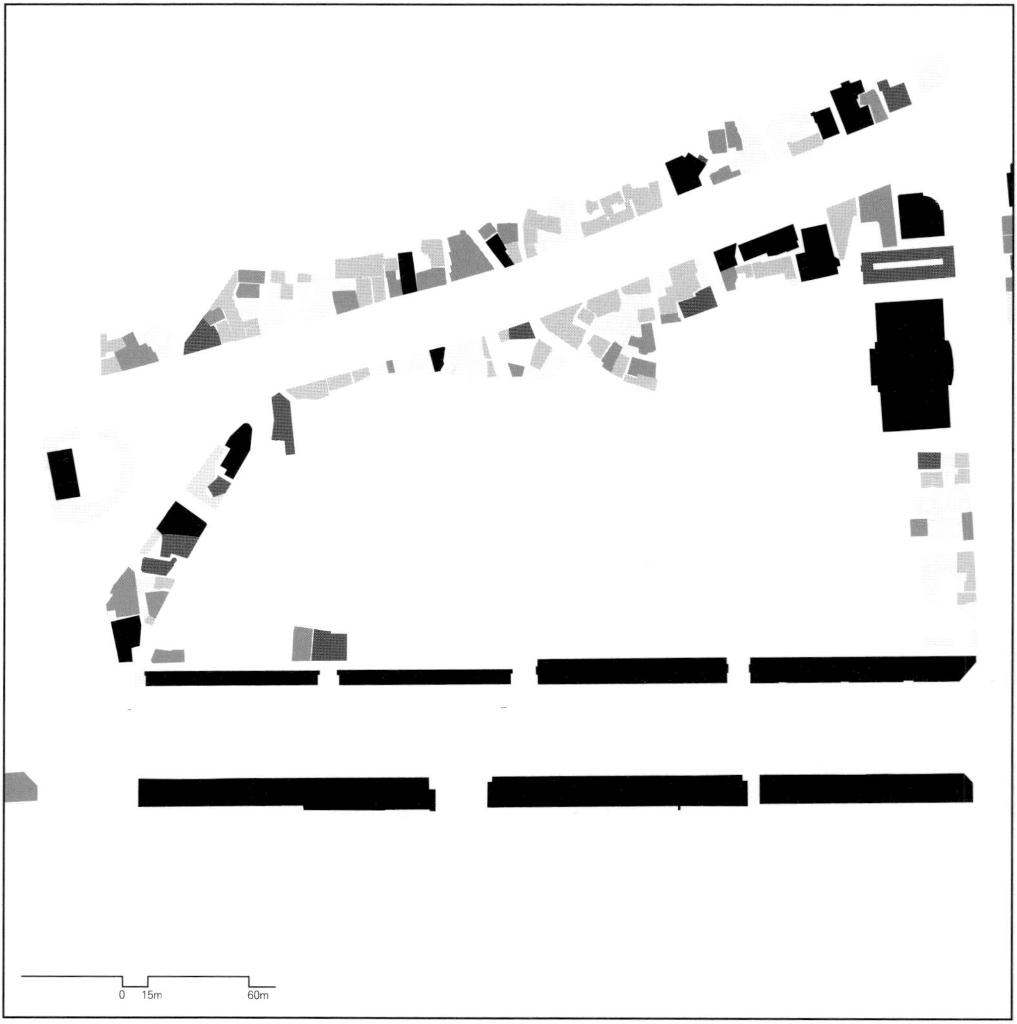

~5.0m　5.0~9.9m　10.0~19.9m　20.0~29.9m　30.0~49.9m　50.0m~

Map 10.

하천: 1958, 1968, 현재

Stream: 1958, 1968, Present

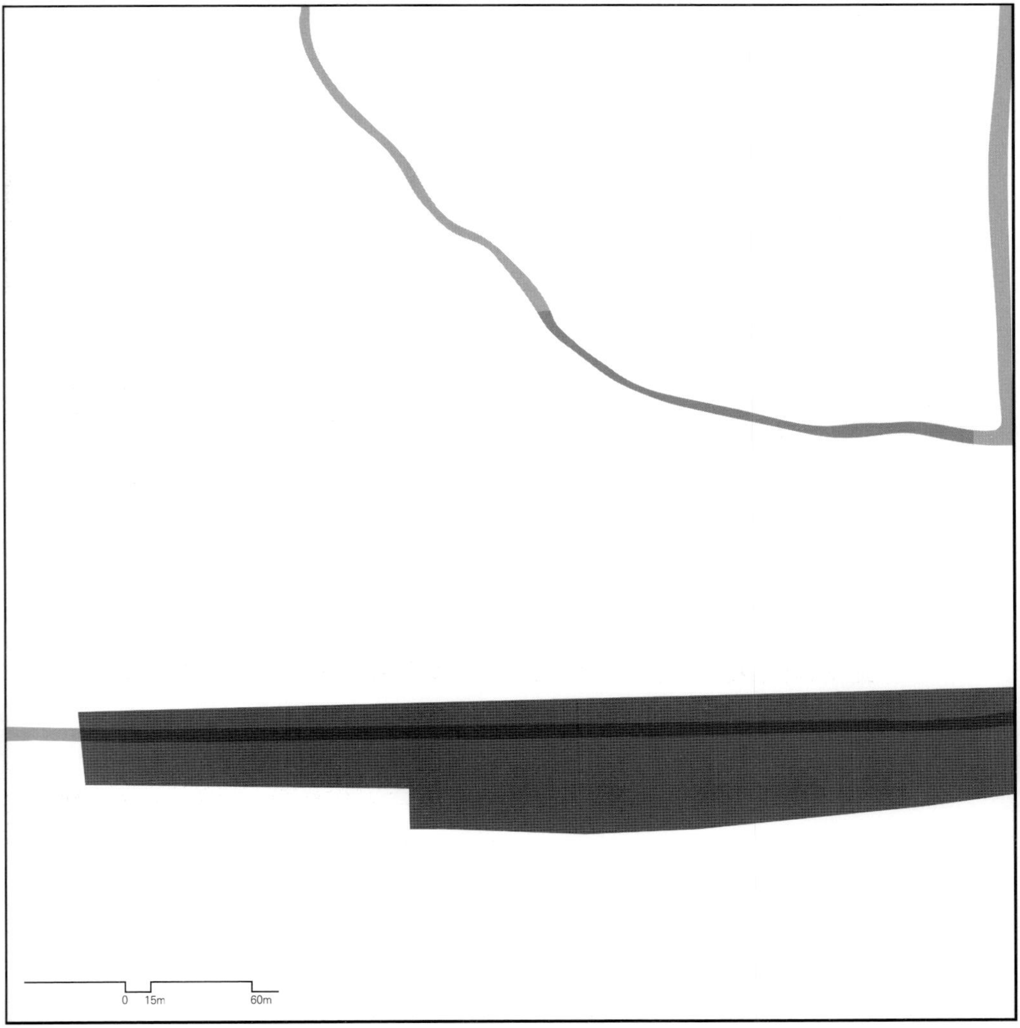

1958 1968 현재 Present

미래

인형: In the Shape of a Human

차세대 첨단 기술 제조업을 위한 동대문 문구완구시장 재편

요란하지 않게 슬그머니, AI의 도움을 받는 시대가 우리 곁으로 다가왔다. 공공의 영역에서 혁명을 일으킬 것이라 여겼던 AI는 오히려 우리 각자의 가정들 속에서 사이보그 시대를 향한 첫걸음마를 뗐다. 아마존 에코Amazon Echo와 애플 홈팟Apple HomePod과 같은 장치들은 공상과학에 자주 등장하는 기술이 지배하는 디스토피아적인 도시의 모습을 훌쩍 건너뛰어 버렸고, 오히려 너무나 쉽게 쓸 수 있고 사용하기 쉽다고 느껴지도록 설계되었다. 흔히 '스마트 스피커'라고 불리는 이 장치들은 기존의 환경에 잘 어우러지도록 요란하지 않은 형태로 제작되지만, 사실 그 속에는 최첨단의 컴퓨터가 포장되어 있다. 이 장치들은 보다 인간처럼 느껴지도록 반드시 우리에게 실시간으로 반응하는데, 이는 달리 말하면 이들이 언제나 '깨어있다'는 뜻이 된다. 과연 이 장치들은 우리의 명령에 즉각 반응하는 하인 같은 것일까? 아니면 우리를 거대한 권력 구조의 노예로 만들어버릴 일방적인 디지털 감시 거울일까?

서울시가 잠정적으로 제정한 '동대문 문구완구시장'이라는 이름으로 불리는 이 황폐한 도시 클러스터는 앞서 말한 가정 기술의 새로운 세대를 이끌어가기 위한 거의 완벽한 무대다. 비록 이 지역의 전근대적인 골목들이 오토바이와 차에 의존하는 현재의 운송 체제에 걸림돌이 되기도 하지만, 그와 동시에 이 지역의 다용도 공간들은 저렴한 임대료를 바탕으로 긴밀하게 협력할 수 있는 독특한 밀도를 지녔다. 게다가 이 지역을 에워싸는 도보 1km 반경 안에는 전파상과 소규모 봉제 공장과 같이 제작에 필요한 모든 '하드웨어'가 속해 있을 뿐만 아니라 변호사, 회계사 그리고 변리사 사무소와 같이 사업체 운영에 필요한 '소프트웨어' 또한 포함된다. 게다가 서울에서 가장 중요한 전통시장과 현대시장들이 바로 인근에 있어 이러한 활동들이 일어나도록 힘을 북돋아 준다.

이 쇠퇴해가는 지역의 생산 가능성을 살리고 더 나은 거주 환경을 만들고자 우리가 제안하는 것은 기존 건물 조직에 도시적 규모의 컨베이어 벨트를 마치 바느질하듯이 엮는 것이다. 이 단순한 기계 시스템은 탈것에 기반을 둔 운송 방식을 대체하여 골목들을 보행자의 것으로 돌려놓는다. 이 지역의 소규모 사업체들은 잘게 분리되어 있지만, 이 컨베이어는 그들이 마치 단일한 대규모 공장처럼 협력적으로 작동할 수 있게 한다. 또한, 이 지역의 역사적인 개울과 생태계는 특수하게 배치된 컨베이어를 통해 제 모습을 되찾아 보행로로 탈바꿈하게 된다. 이들은 이곳이 지닌 깊은 역사와 임박해오는 첨단 기술의 미래 사이에 하나의 신경망을 형성한다.

동대문 문구완구시장의 잠재력에 대해 곰곰이 생각해볼 수 있게 하려면, 우리는 사실에 기반을 두어 '스마트 돌(doll)' 또는 인형의 발명에 관한 상상의 이야기를 풀어나가기로 했고, 그를 위한 소재로 한국에서 가장 인기 있는 매체인 '웹툰'을 사용했다. 이 이야기 속에서 더 발전된 하드웨어는 아이들이 가지고 노는 인형의 부드러운 천 소재로 감싸져 있고 이것이 바로 '스마트 스피커'의 새로운 버전이 된다. 이러한 AI의 제작이 가능한 것은 창신동의 봉제 공장, 세운 상가의 전파상 그리고 을지로의 3D 프린트 제작업소라는 세 업종이 밀집해 있는 이 지역만의 특수성 덕분이다. 이 우화의 핵심은 영어 단어 돌(doll)과 우리말 인형 사이의 미묘하면서도 중요한 차이점에 있다. 전자가 작은 모형 조각상을 물건 취급하고 객관화하는 느낌이지만 인형이라는 단어는 어떤 존재의 상태를 나타내는 말로, 직역하면 "인간의 형상을 한" 상태를 말한다.

인형, 암호명 데이지DAISY: Daycare Artificial Intelligence SYstem의 잠재력은 현재의 모니터링 시스템보다 한참 앞서있다. 데이지는 집의 자동화 시스템과 도시의 감시 네트워크에 완전하게 연결되어 마치 친구처럼 아이의 안전과 건강을 관리하고 관찰한다. 데이지는 창신동 지역의 소규모 공장들에서 대량으로 주문 제작하기 때문에 업무로 바쁜 부모들로 구성된 소비자가 늘어남에 따라 계속 발전해 나간다. 한층 더 높은 감각의 기술을 향하여…

그러나 이 프로젝트가 제안하는 도시재생의 이면에 웅크린 낙관주의에 대한 중요한 질문을 놓칠 수는 없다. 아무리 지역 소비자를 위한 지역 기반의 독립적인 산업체에서 개발된 기술이 대기업을 피해 갈 수 있다고 하더라도, 정말 우리는 방대하게 연결된 권력 구조의 그물에서 벗어날 수 있는 것일까? 아마존은 급속도로 발전하는 그들의 AI 상품에 대해 "유효하고 법적 구속력이 있는 합법적인 요구 없이는" 절대로 소비자의 사적 정보를 이용하거나 노출하지 않을 것이라 장담한다. 그러나 대체 여기서 말하는 요구의 범위는 무엇일까? 특정인(또는 악성 소프트웨어의 주체)이 우리의 사적 정보에 접근하여 악의적으로 이용하는 일은 정말 없는 것일까? 대량으로 주문 제작하는 전자제품이 안전과 편의를 약속하고 있는 와중에, 우리의 이번 작업은 지나칠 정도로 많은 것이 연결된 우리 시대에서 사생활과 민주주의의 의미에 대해 더욱 높은 차원의 질문을 던진다.

Future

Inhyeong: In the Shape of a Human

Retooling the Dongdaemun Toy and Stationery Market for the Next Generation of High-Tech Manufacturing

The age of AI assistants is here, "not with a bang, but with a whimper." Rather than a revolution within our public sphere, the first baby steps toward our cyborg future are taking place within our homes. Leap-frogging over sci-fi tropes that paint the city as a technologically dominated dystopia, devices such as the Amazon Echo and Apple HomePod are exceedingly easy to use and are designed to appear that way: Innocuously called 'smart speakers,' they are actually state-of-the-art computers packaged in unobtrusive bodies that blend into our existing environments. To be more human, they must interact with us in real time, which also means they are always 'awake.' Are they our responsive servants, or are they digital one-way surveillance mirrors putting us in servitude to larger power structures?

A dilapidated urban cluster just east of Seoul's historical core, provisionally named by the city as the Dongdaemun Stationary and Toy Market (DDMSTM), is almost perfectly suited to stage the production of this next generation of domestic technology. Even as the area's pre-modern alleyways are an impediment to current delivery logistics that rely on mopeds and cars, its unique density is capable of harboring intensely collaborative mixed-use at low rents. Moreover, a one kilometer walking radius surrounding the area includes all the 'hardware' for fabrication, such as electronics and garment micro-factories, as well as the 'software' for running businesses, such as lawyers, accountants, and patent offices. Bolstering this activity are many of Seoul's most important traditional and modern markets in direct vicinity.

To invigorate the productive capacity of this declining area and create a better residential environment, our proposal inserts an urban scale conveyor belt, like a needle and thread, through the existing building fabric. This simple mechanical system replaces the need for vehicular-based delivery, releasing the alleyways back to pedestrian use. Even as the area's micro-businesses are atomized, the conveyor allows them to act collectively as one mega-factory. Through the conveyor's specific configuration, a historical stream and ecosystem is uncovered and transformed into a pedestrian path, creating a synaptic relationship between the region's rich history and its imminent high-tech future.

To ruminate about the DDMSTM's potential, we use the popular medium of Korean 'webtoons' to track a fact-based fictional narrative of the invention of a 'smart doll' or inhyoung. As the next iteration of the 'smart speaker,' advanced hardware is embedded within the soft fabric of a child's doll. The unique convergence of three local capabilities make the production of this AI possible: Changsindong's garment factories, Sewoon Arcade's electronic markets, and Euljiro's 3D fabrication shops. Central to the allegory is the subtle, but important difference between the English word doll and the Korean equivalent, inhyoung. While the former objectifies a model figurine, inhyoung is a state of being, directly translated as, "in the shape of a human."

The potential for inhyoung, code-named DAISY (Daycare Artificial Intelligence SYstem) is far beyond current monitoring systems. It acts as a surrogate friend watching over a child's safety and health, seamlessly interfacing with the home's automation system and the city's surveillance network. Because DAISY is mass-customized by the micro-factories within our site, an ever-growing buyer-base of overworked parents push its development to higher levels of sentience…

A core question haunts the optimism behind this project's proposed urban regeneration: Even if we bypass the big brands with technology created by independent local businesses for local customers, can we really be separated from the power structures of the vastly connected web? In regards to its rapidly growing AI products, Amazon claims that it will not release private customer information "without a valid and binding legal demand." What is the threshold for this demand? And can a person (or malware entity) gain access and maliciously exploit our private information? Even as safety and convenience is promised by mass-customized electronic products, our proposal probes the meta-questions concerning privacy and democracy in our emerging hyper-connected era.

Cartoon

DAISY: Daycare Artificial Intelience SYstem

Project : Architecture

Drawn by Hyein Kim

*전시 「도시생활주거: Factory for Urban Living」를 재구성한 것으로, 국문 번역본이 수록되지 않았습니다.

Sunday 6pm, Isaac' s House
Daddy, Daisy and I want to play!
uhh.. what will I say...
...you look so pretty!
Daisy doesn't speak like that...
These were the magical words, and Suzie knew that using it wouldn't let me refuse her request.
Mommy was really good at it...

MONDAY 8:51 AM, ISAAC'S HOUSE
... I'LL HAVE ENOUGH TIME UNTIL SHE GOES BACK FROM KINDERGARTEN.
9:36 AM, HWANG HAK-DONG FLEA MARKET

HMMM...
THIS? OR THAT...
DO YOU HAVE SOMETHING TO BUY?
YEAH, IS THIS THE SMALLEST RECORDER YOU SELL?
OLD MAN
MAYBE? IT IS ONLY 3,000 WON BUY IT!
WELL ... OKAY!
1:08 PM, ISAAC'S HOUSE
HMM
HMHMMM
5:09 PM, ISAAC'S HOUSE
DADDY, I'M HO...
WHAT ARE YOU DOING?

GOOD AFTERNOON SWEETIE! WANNA SEE?
YES, SHOW ME!
CLICK!
YOU LOOK SO PRETTY-
WOW- CAN I PLAY WITH HER NOW?
UNTIL THEN, I DID NOT KNOW HOW HARD IT WAS TO PLAY WITH A FIVE-YEAR-OLD GIRL.
LET'S EAT DINNER FIRST, AND THEN WE CAN PLAY WITH DAISY.
LET'S GO TO A PARTY!
I LOVE IT.
YOU LOOK SO PRETTY-
6:30 PM
I LOVE IT.
I LOVE IT.
LET'S GO TO A PARTY!
7:30 PM
LET'S GO TO A PARTY!
LET'S GOTO A PARTY!
LET'S GO TO A PARTY!
8:30 PM
DADDY... ISN'T IT FUN TO PLAY WITH DOLLS WITH ME?
WHY DON'T YOU MOVE THE DOLL LIKE I DO?
IT IS FUN!
...JUST DADDY'S ARM ACHES...

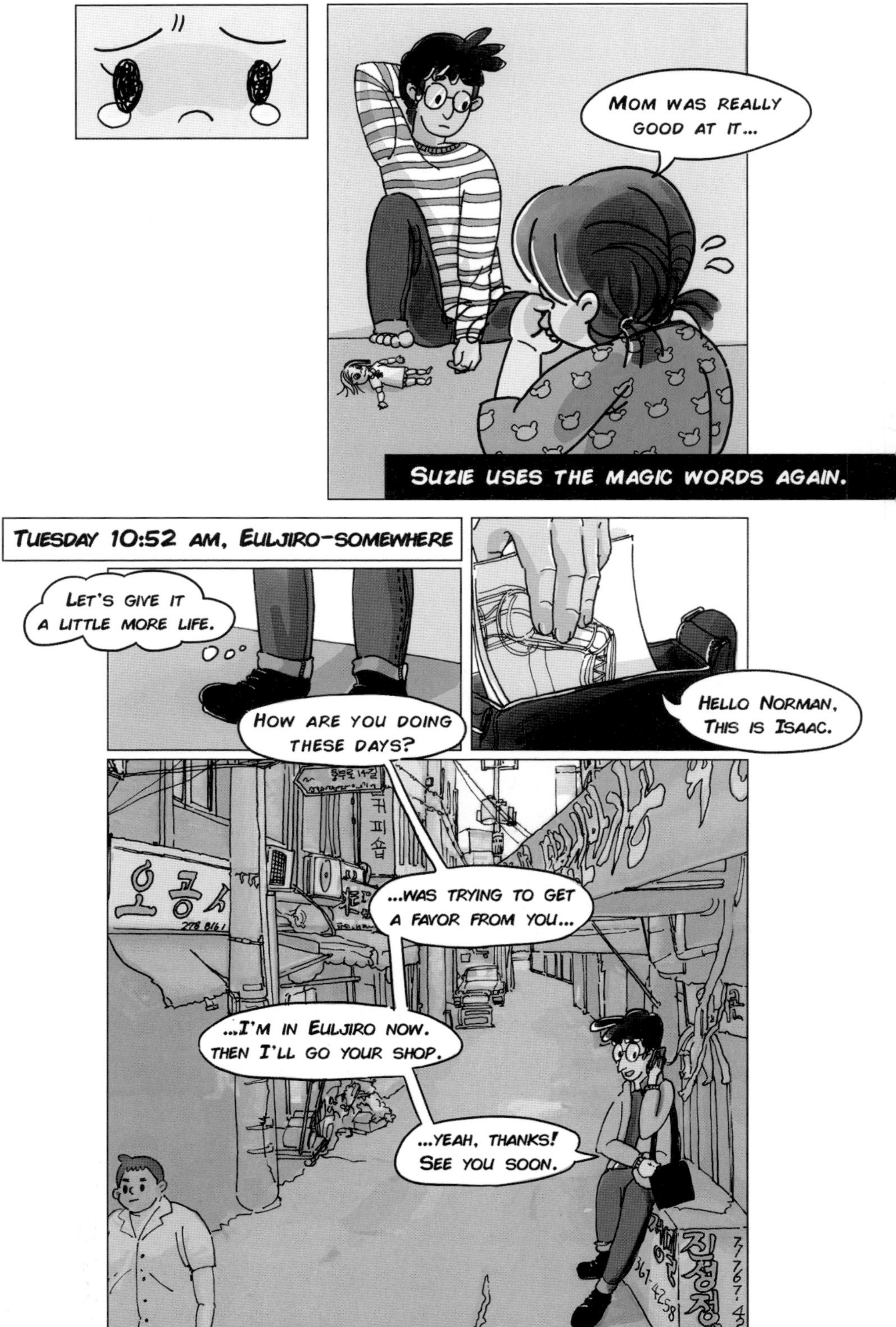

MOM WAS REALLY GOOD AT IT...
SUZIE USES THE MAGIC WORDS AGAIN.
TUESDAY 10:52 AM, EULJIRO-SOMEWHERE
LET'S GIVE IT A LITTLE MORE LIFE.
HELLO NORMAN, THIS IS ISAAC.
HOW ARE YOU DOING THESE DAYS?
...WAS TRYING TO GET A FAVOR FROM YOU...
...I'M IN EULJIRO NOW. THEN I'LL GO YOUR SHOP.
...YEAH, THANKS! SEE YOU SOON.

3 DAYS LATER
FRIDAY, 3:27 PM, ISAAC'S HOUSE
DING DONG—
WAIT A SECOND~!
HERE IS THE PAYMENT, KEEP THE CHANGE, THANK YOU.
UICK SERVICE.
NANANNA~
공업사
NANANA~
SCREECH—
BAM!

5:03 PM, ISAAC'S HOUSE
DAD, I'M HO...
SUZIE, LOOK HERE!
HUH-?
HERE, SWEETIE.
?
LET'S GO TO A PARTY!
DADDY, YOU ARE THE BEST! I LOVE YOU DADDY!
A FEW YEARS LATER: LECTURE HALL
AFTER DAYS OF NO SLEEP, MY EYES WERE RED, MY FACE WAS PALE...
BUT AFTER SEEING MY DAUGHTER'S FACE, I COULD FINALLY SMILE.
IF I STOPPED THERE, DAISY WOULD HAVE...
...JUST COULD HAVE BEEN FINISHED AS MY DAUGHTER'S HIGH-END TOY.
A FEW YEARS BEFORE, ISAAC'S HOUSE

Daddy, I need more dolls like Daisy.
What? Is it not enough for you?
No... I mean, my friends Sera and Anna want to have Daisy.
But I don't want to give her mine...
At that time I realized, Oh, Daisy can be a friend of many children like my daughter!
Lecture hall
I started to move aggressively so that more kids could meet the doll.
I invited designers to create a look
Sindang Creative arcade
that embodies Daisy's identity.

THE MOST MEANINGFUL MOMENT AFTER SETTING UP THE CENTER IS WHEN USERS CAME RIGHT HERE TO ASK FOR THEIR DESIRED FUNCTIONALITY.

BEGINNING WITH THAT,
MANY FEATURES HAVE BEEN ADDED TO DAISY.

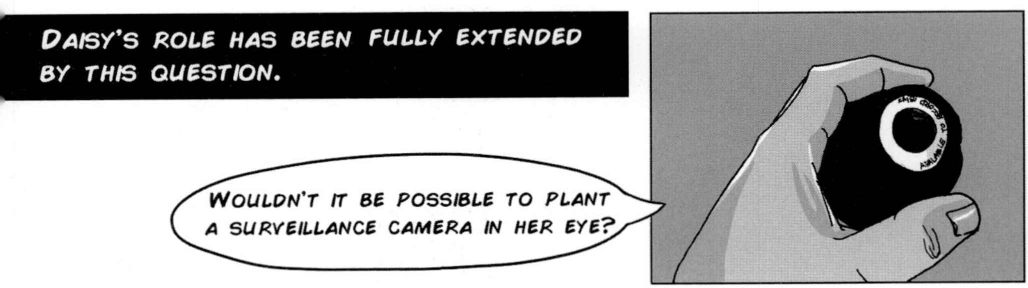

Ad running
Daycare Artificial Intelligence SYstem
DAISY®
DAISY® LOOKS AT MY SON AND SHOWS ME WHAT MY SON DOING!
DAISY
DAISY® TALKS TO MY SON AND TELLS ME WHAT HAPPEND TO HIM!
!!!
DAISY
WHEN DAISY® SENSES AN EMERGENCY SITUATION REGARDING MY SON...
DAIS
SHE WORKS EVEN IF SHE IS IN SLEEP MODE!
BEEP—! BEEP—!
DAIS
DAISY® WANTS TO SHARE HER EVERYDAY LIFE WITH US!
* ON AIR *
MISS APPLE
STAR
CONNECT YOUR LIFE. OO TELECOM

광고 상영중
우리집 인공 지능 매일육아 도우미
데이지®
데이지®는 아이를 관찰하며
저에게 아들이 무엇을 하는지 보여줘요!
DAISY
데이지®는 아이와 이야기하며
아들에게 무슨 일이 있었는지 들려줘요!
!!!
DAISY
아들에게 어떤 응급상황이 생겼을 때
데이지®는 그것을 감지하고
DAIS
스스로 작동해요!
절전 모드에서도 가능하답니다!
BEEP—!
BEEP—!
DAIS
데이지®는 일상생활을
항상 우리와 함께 나누고 싶대요!
ON AIR
MISS
APPLE
STAR
당신의 삶을 연결하다, OO 텔레콤

LECTURE HALL
THROUGH ALLIANCE WITH OO TELECOM, I WAS ABLE TO USE THEIR ABUNDANT CAPITAL TO EXPAND MY BUSINESS...
WITH THE CONDITION OF APPLYING THEIR IOT SYSTEM TO DAISY.
IN THE LAST FEW YEARS, DAISY'S PRODUCTION HAS RISEN SHARPLY.
SELL
TIME
BUT LOCAL BUSINESSES ARE ESSENTIAL TO OUR TEAM
DAISY
DAISY
SO WE WERE ABLE TO REVITALIZE THIS PLACE AGAIN.

DONGDAEMUN STATIONERY AND TOY MARKET
빵파레본점
979-1452-3
완구
3-8962
200m

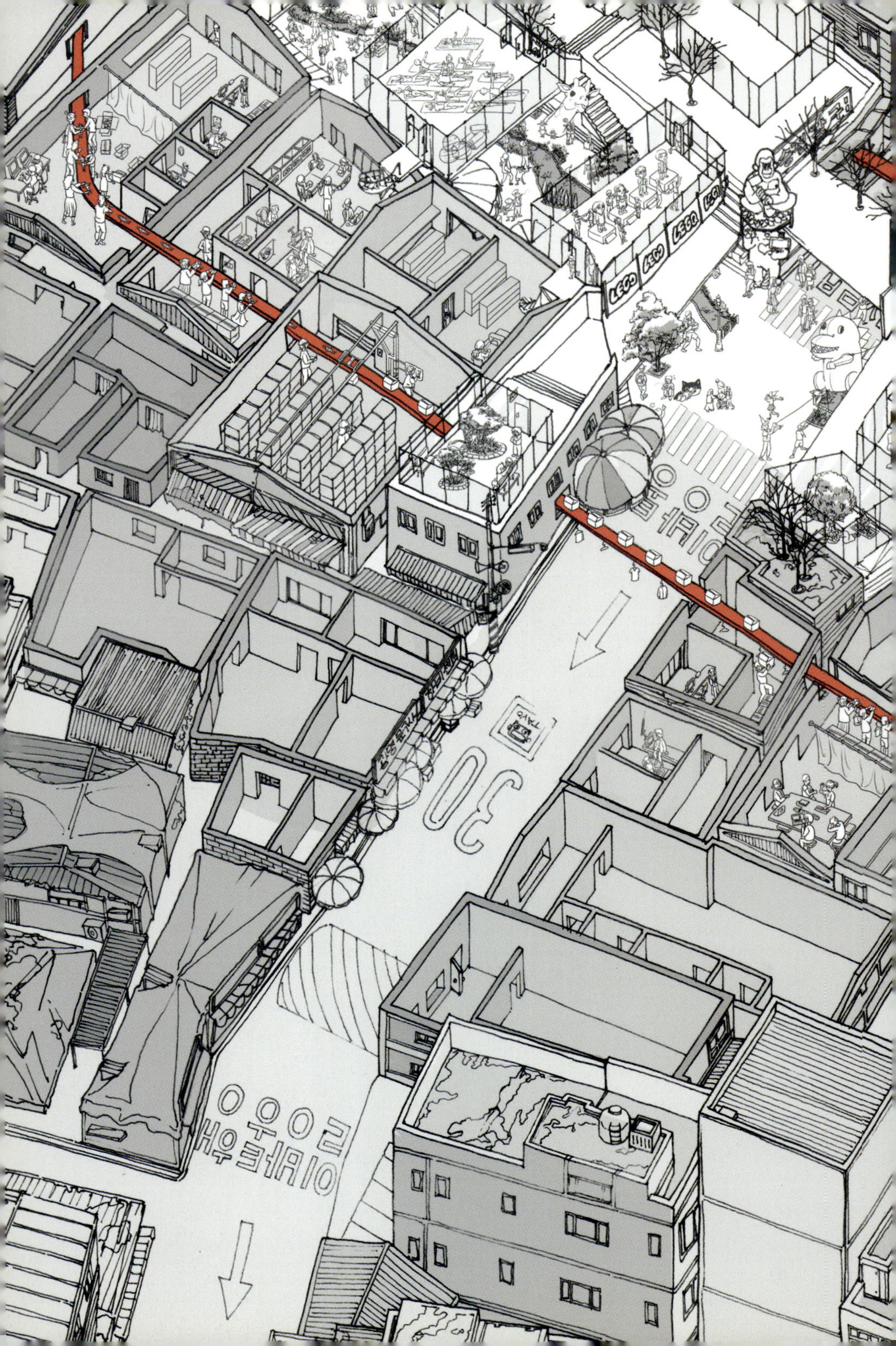
LEGO
30

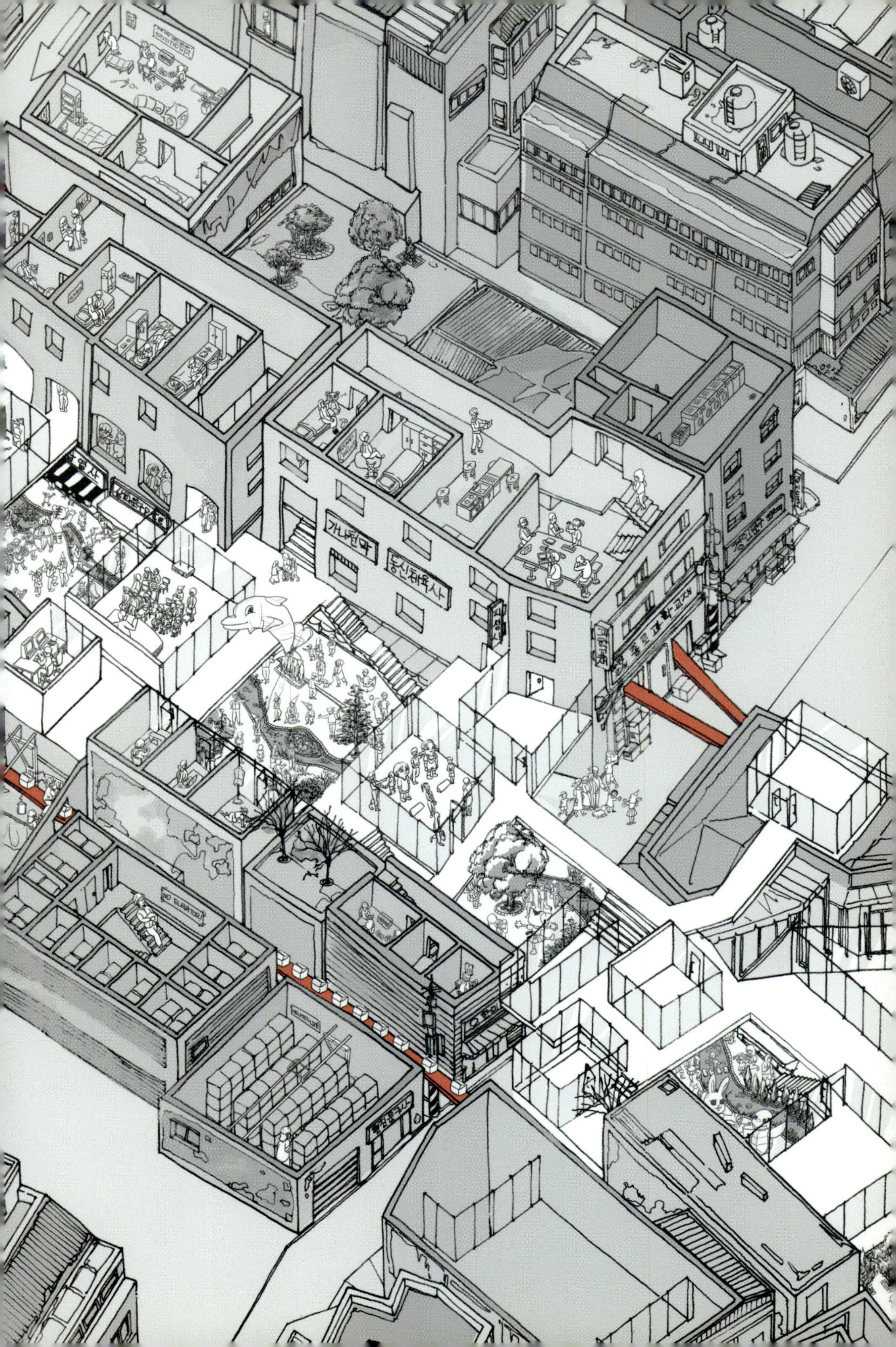
동신체육사
NO ELEVATOR

LECTURE HALL
THE DOLLS THAT I MADE FOR CONNECTING MY DAUGHTER AND ME,
...HAS CONNECTED KOREAN PARENTS AND CHILDREN,
AND NOW BEGINS TO TO CONNECT THE PARENTS AND CHILDREN OF OTHER COUNTRIES.
NOW IS THE TIME TO TELL YOU, YOUR FAMILY, AND DAISY.
THANK YOU! THANKS FOR EVERYONE!
THIS WAS THE STORY OF DAISY AND ME, AND MY DAUGHTER.

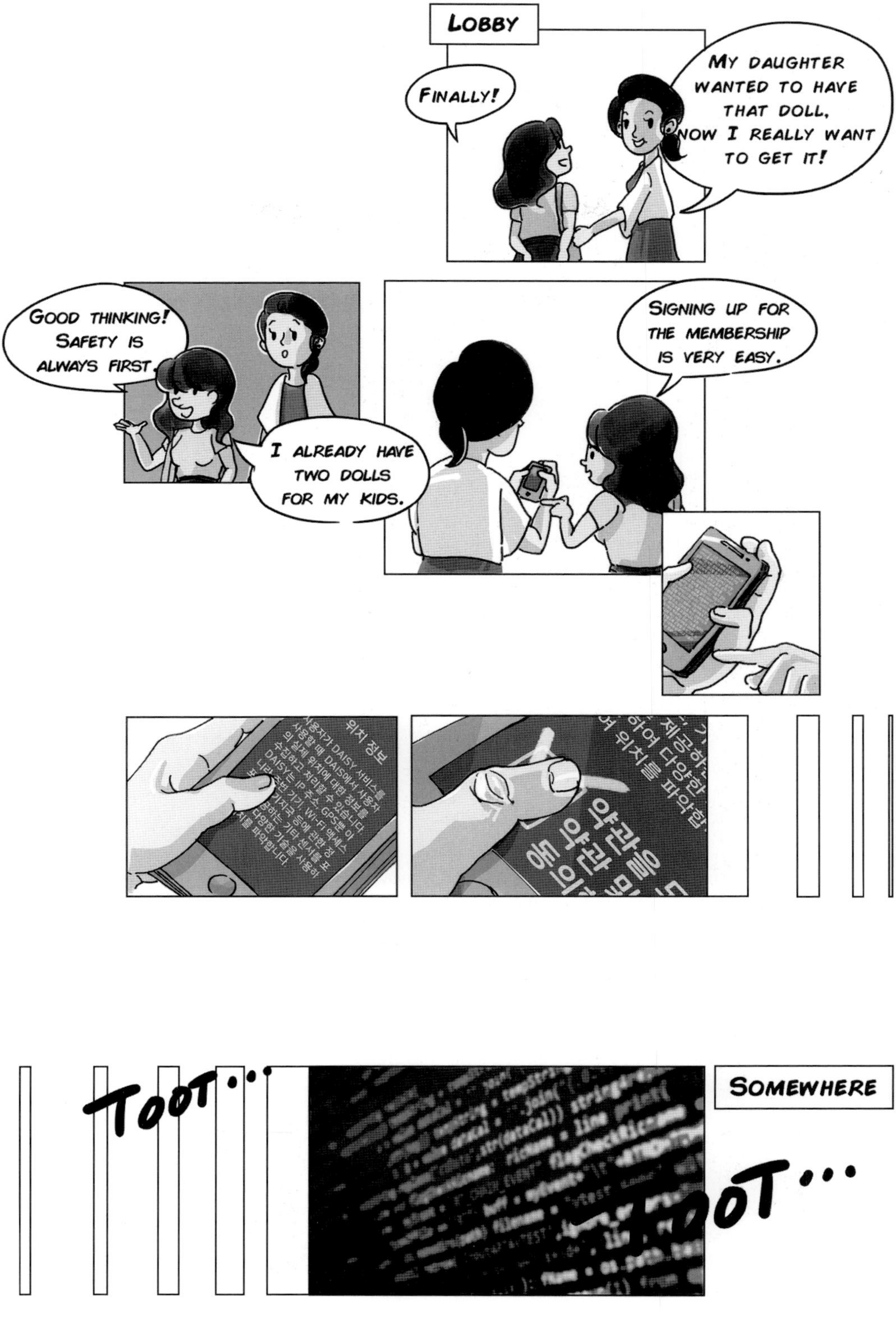
LOBBY
FINALLY!
MY DAUGHTER WANTED TO HAVE THAT DOLL, NOW I REALLY WANT TO GET IT!
GOOD THINKING! SAFETY IS ALWAYS FIRST.
I ALREADY HAVE TWO DOLLS FOR MY KIDS.
SIGNING UP FOR THE MEMBERSHIP IS VERY EASY.
TOOT...
SOMEWHERE
TOOT...

TK-
TK-

EVERYONE IS OK TONIGHT.

4TH INDUSTRIAL REVOLUTION

:Urban production and mass customization

To re-imagine the existing capabilities in the area, a fact-based fictional cartoon traces the production of DAISY (Daycare Artificial Intelligence System), a mass-customized toy that watches over children's' safety. The story also sends a warning about urban regeneration and the potential takeover by corporate interests.

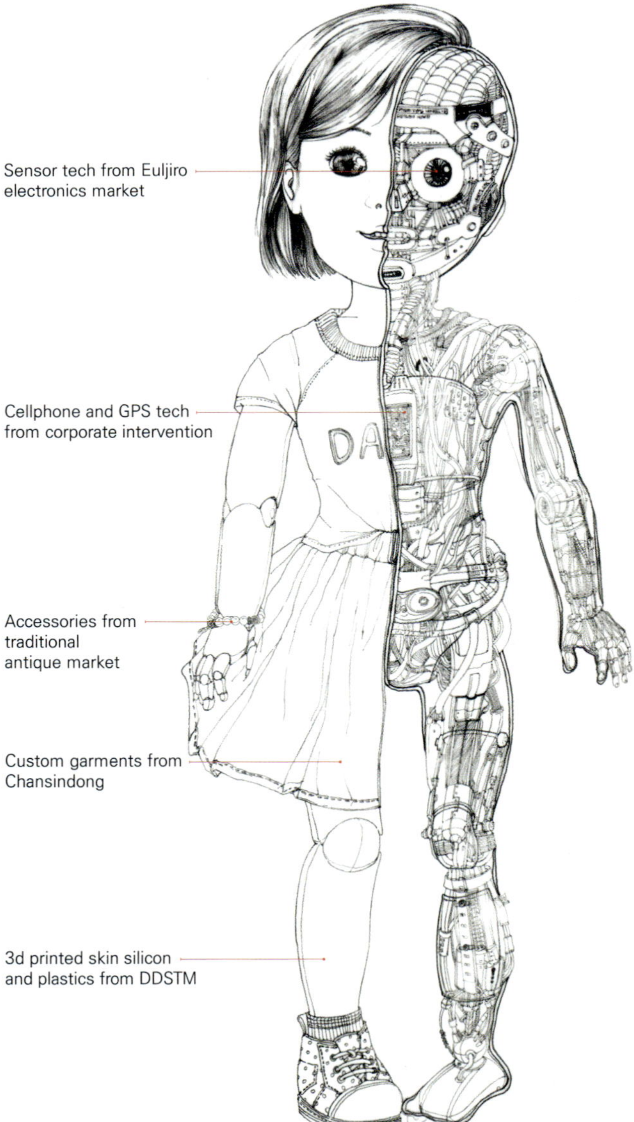

CONVEYOR BELT TYPOLOGIES

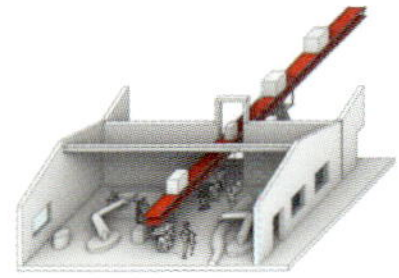

through existing factories

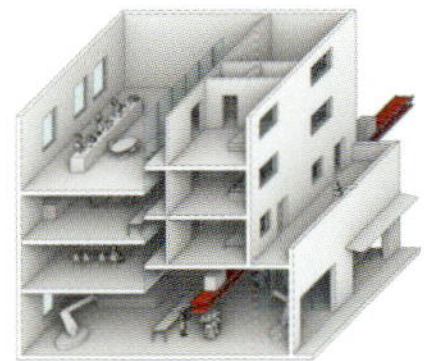

between factories and retail buildings

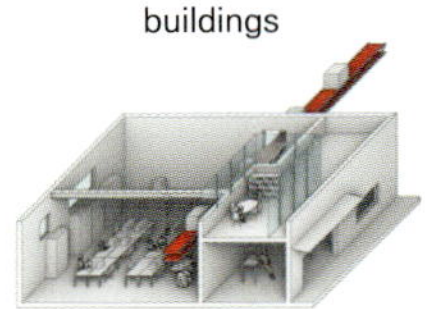

behind retail back of house

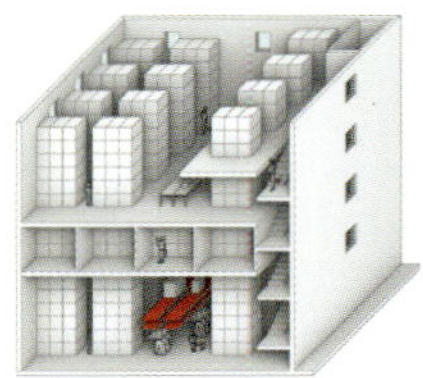

inside storage warehouses

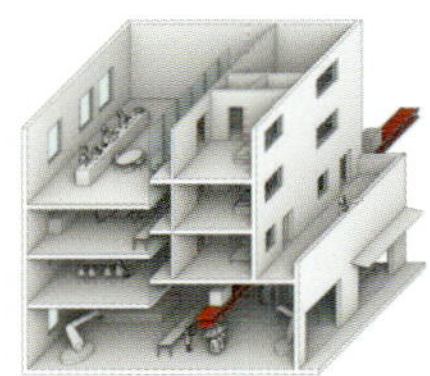

under mixed-use live-work

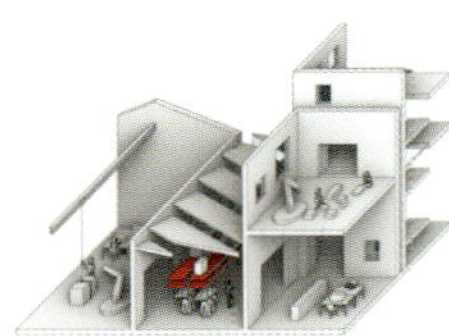

serving a production arcade

PROPOSED URBAN INTERVENTION

The urban-scale conveyor belt allows the diverse programs to collaborate as one urban system

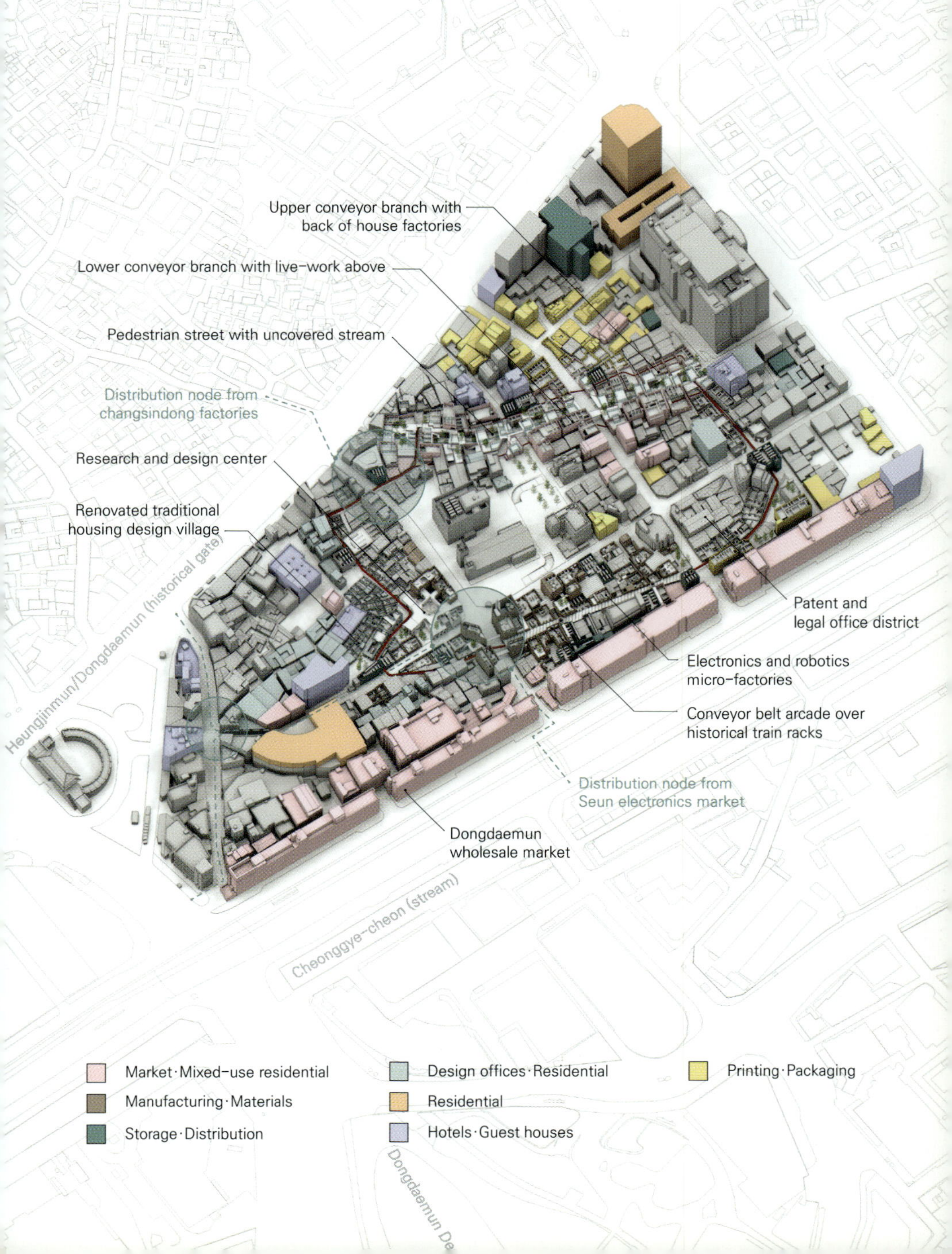

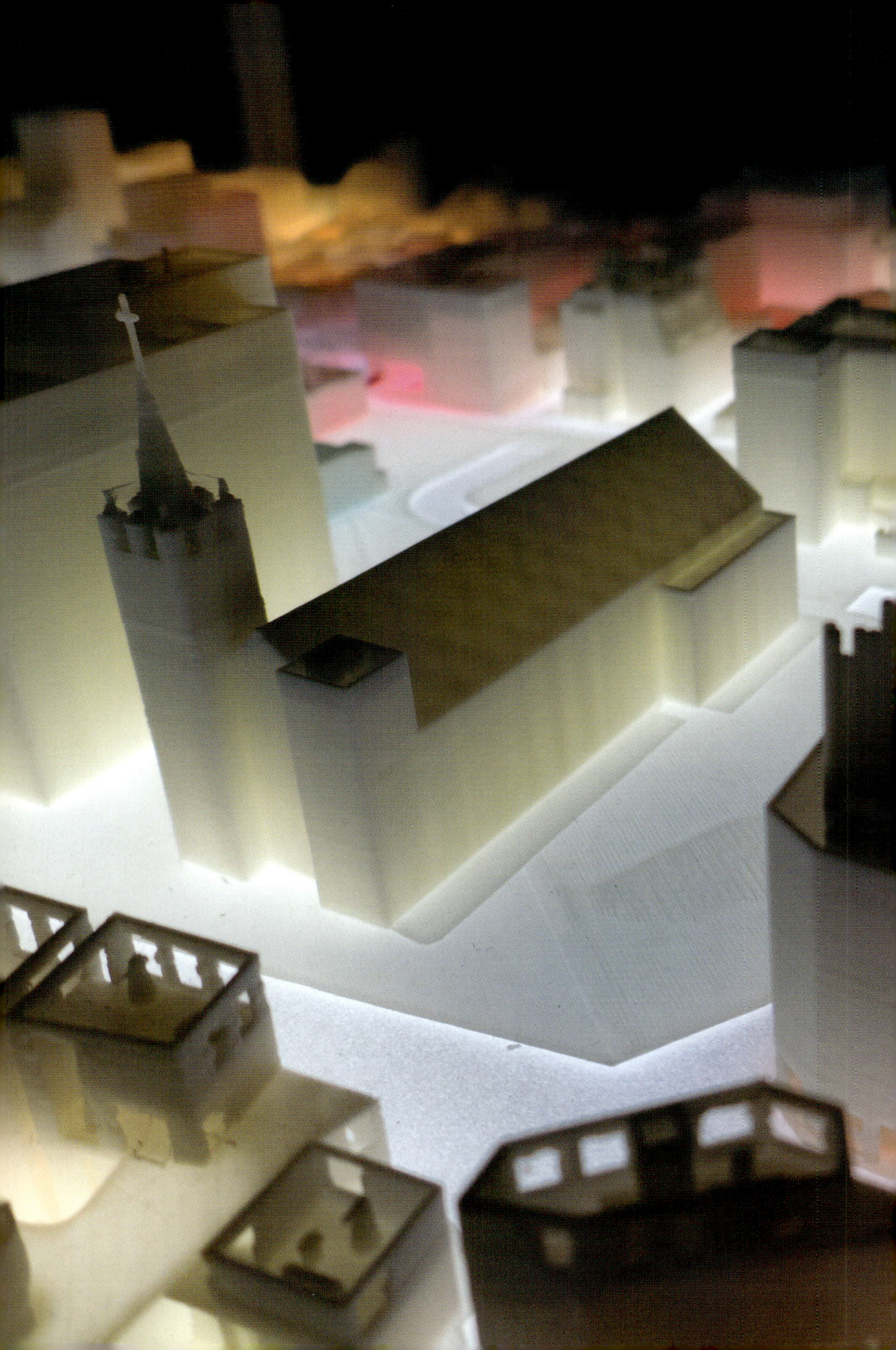

철도에서 첨단 기술 배송으로 — 대상지의 남쪽에 있는 역사적인 철도는 초소형 공장 사이의 협업과 첨단 기술 제작을 촉진하기 위한 컨베이어 벨트로 바뀐다.

From Railroad to Hightech Delivery — The historical railroad on the south side of the site is converted into a covered conveyor to expedite hightech fabrication between collaborating micro-factories.

도시 하천 — 컨베이어는 문구완구상점의 유휴공간을 활용하여 차량이 주요 도로를 통행할 필요성을 제거한다. 이를 통해 주요 도로의 역사적인 하천을 복원하고 새로운 프로그램이 담긴 다리를 설치할 수 있기 때문에 밀도 높고 복합적인 활용이 가능해진다.

Urban Stream — The conveyor services the back-of house of the Toy Market removing the need for vehicles along the main street. This allows the uncovering of a historical stream with new program bridges that bring density and mixed-use.

전통 마을 × 새로운 디자인 허브 — 새로운 디자인 허브는 지역 사회의 모임 장소이자 전통 '한옥' 마을의 관문이 된다.

Traditional Village × New Design Hub — A new design hub becomes a community meeting place and gateway to the traditional 'hanok' village.

대안적인 미래들

서울대학교 4-1 스튜디오: 단면도를 통해 다시 구상하는 도시재생

단면도, 특히 도시의 밀도를 다루는 단면도는 여러 용도가 혼합된 도시의 잠재력을 드러내고, 구체적이면서도 유형학적인 방식으로 소통하는 데에 중요한 것들을 드러낸다. 그러나 평면도를 사용할 때의 장점이 결여된 섹션만으로는 쉽게 그 효용성, 즉, 도시 공간에 어떤 영향을 미치는가에 대해서는 알아차리기 힘들다. 그러나 투시 단면도의 형태를 취하여 삼차원을 추가한 이 복합적인 도면composite drawing은 도시 맥락 속에 쉽게 녹아들어, 도시의 순환 체계 그리고 특수한 사건들과 긴밀하게 연결된다. 삼차원을 표현하는 방식을 통해, 단지 이차원의 매체만으로도 시간이라는 사차원의 개념 또한 도면과 관계를 맺고, 도면을 통해 활성화된다.

서울대학교의 이번 4학년 스튜디오는 새로운 프로그램과 공간을 제안하는 주 도구로 투시 단면도를 사용하며, '도시 재생'의 규모에서 동대문 문구완구시장에 잠재된 가능성의 문을 두드린다. 이들 단면도는 도시 설계, 열린 공간, 기존의 구조물 그리고 실내 건축에 이르는 다양한 스케일을 종합하여 창작의 새로운 지평을 연다. 이번 스튜디오에서는 이 책의 서문에 등장하는 '버려진 땅(Terrain Vague)'이라는 개념 또한 다루어졌는데, 이는 근대주의자들이 기대하는 프로그램과 경제 논리로서의 '생산성'이라는 헤게모니에서 벗어난 도시의 '제3 공간'을 의미한다. 예를 들면 이 지역의 거주 공간 중 다수는 다양한 용도가 섞여 있는 도시에 필요한 문지방 영역이 된다. 이 틈새 공간의 작동 방식은 프로그램들을 모호하게 분리하여 한때 전용함으로써 도시에 고도의 다양성이 존재할 수 있게 한다.

의류 디자인 클러스터
박상준
10
인스턴트 도시
최수연
01
어울리는 동네
한경태
11
어린이 교육 센터
문서령
02
문지방 공간
다시 쓰기
카밀라 포레로 사무디오
12
깊숙한 표면
김동환
03
지면 곱하기 지면
김창용
09
한데 드러난 시간
홍수화
04
창신 시장 캠퍼스
김상범
08
연결된 플랫폼
박재성
05
씨실날실 클러스터
김동현
07
제목 없음
키에란 맥과이어
06

Alternative Futures

SNU Studio 4-1: Reimagining Urban Renovation Through the Section

The section drawing, especially within the density of the city, can reveal the potential of mixed-use urbanism. It distills relationships in ways that communicate both specifically and typologically. But without the benefit of the plan, we cannot discern the efficacy of the section: how it enacts its influence in the urban field. By adding the 3rd dimension, however, in the form of a *perspective* section, this composite drawing is suddenly located within its context, entangled with circulatory systems and contingencies. Through the representation of the 3rd dimension, the 4th dimension of time is referenced and activated, all within a 2-dimensional medium.

The 4th-year studio at Seoul National University worked across the scales of 'urban renovation' using the perspective section as its main tool to propose ways in which new programs and spaces tap the latent potential of the DDMSTM region. The sections coordinate the scales of urban design, open space, existing structures, and interior architecture to spark new conditions of creativity. Moreover, the idea of the 'Terrain Vague' mentioned in this book's preface as urban '3rd space' outside the hegemony of modernist expectations of 'productivity' is also addressed: The site's many residual spaces become liminal zones essential to mixed-use urbanism. These gaps allow heightened diversity through the way they ambiguously separate programs allowing them to be momentarily appropriated.

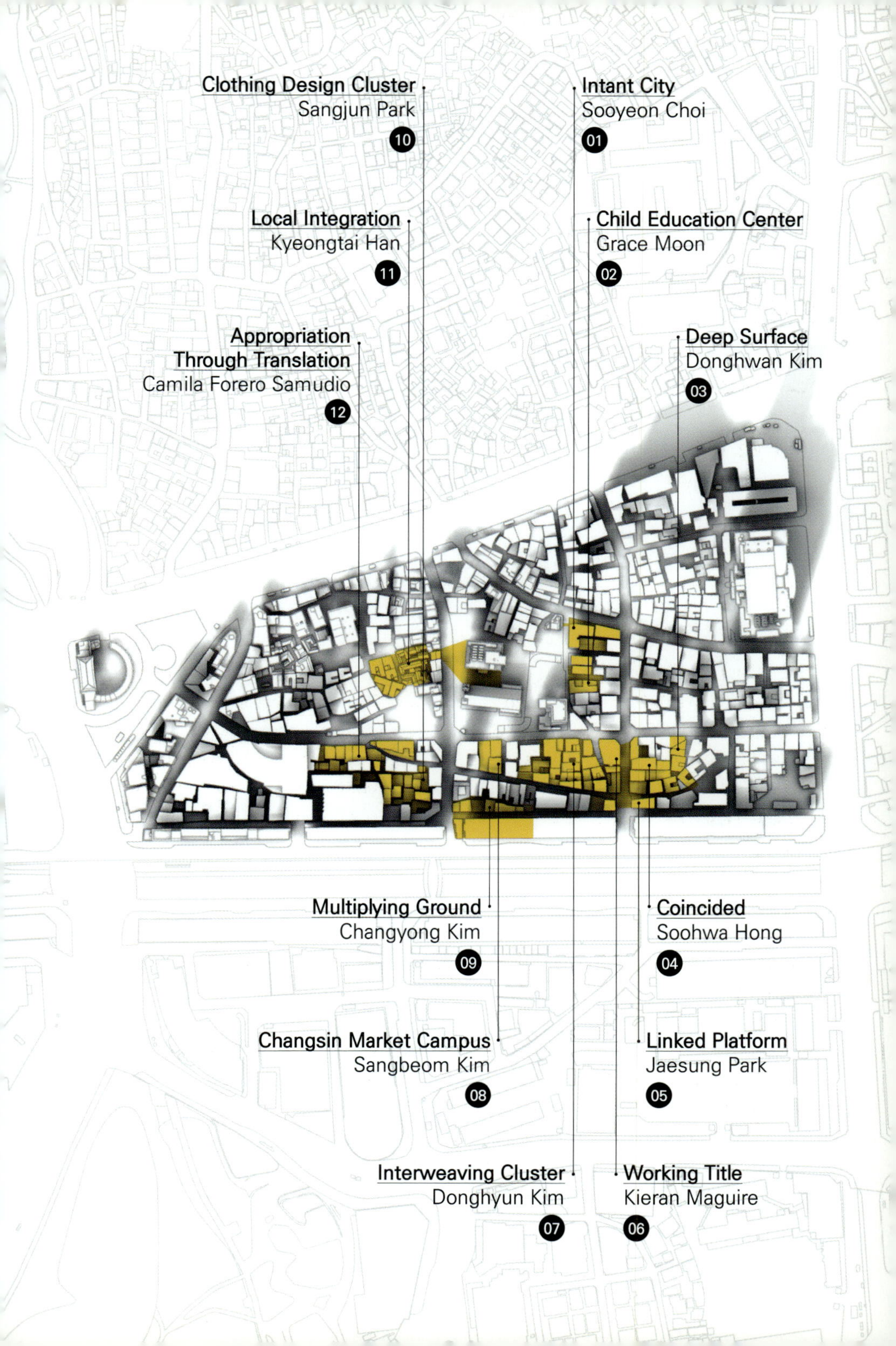
Clothing Design Cluster
Sangjun Park
10
Intant City
Sooyeon Choi
01
Local Integration
Kyeongtai Han
11
Child Education Center
Grace Moon
02
Appropriation
Through Translation
Camila Forero Samudio
12
Deep Surface
Donghwan Kim
03
Multiplying Ground
Changyong Kim
09
Coincided
Soohwa Hong
04
Changsin Market Campus
Sangbeom Kim
08
Linked Platform
Jaesung Park
05
Interweaving Cluster
Donghyun Kim
07
Working Title
Kieran Maguire
06

인스턴트 도시

INSTANT CITY: SOOYEON CHOI

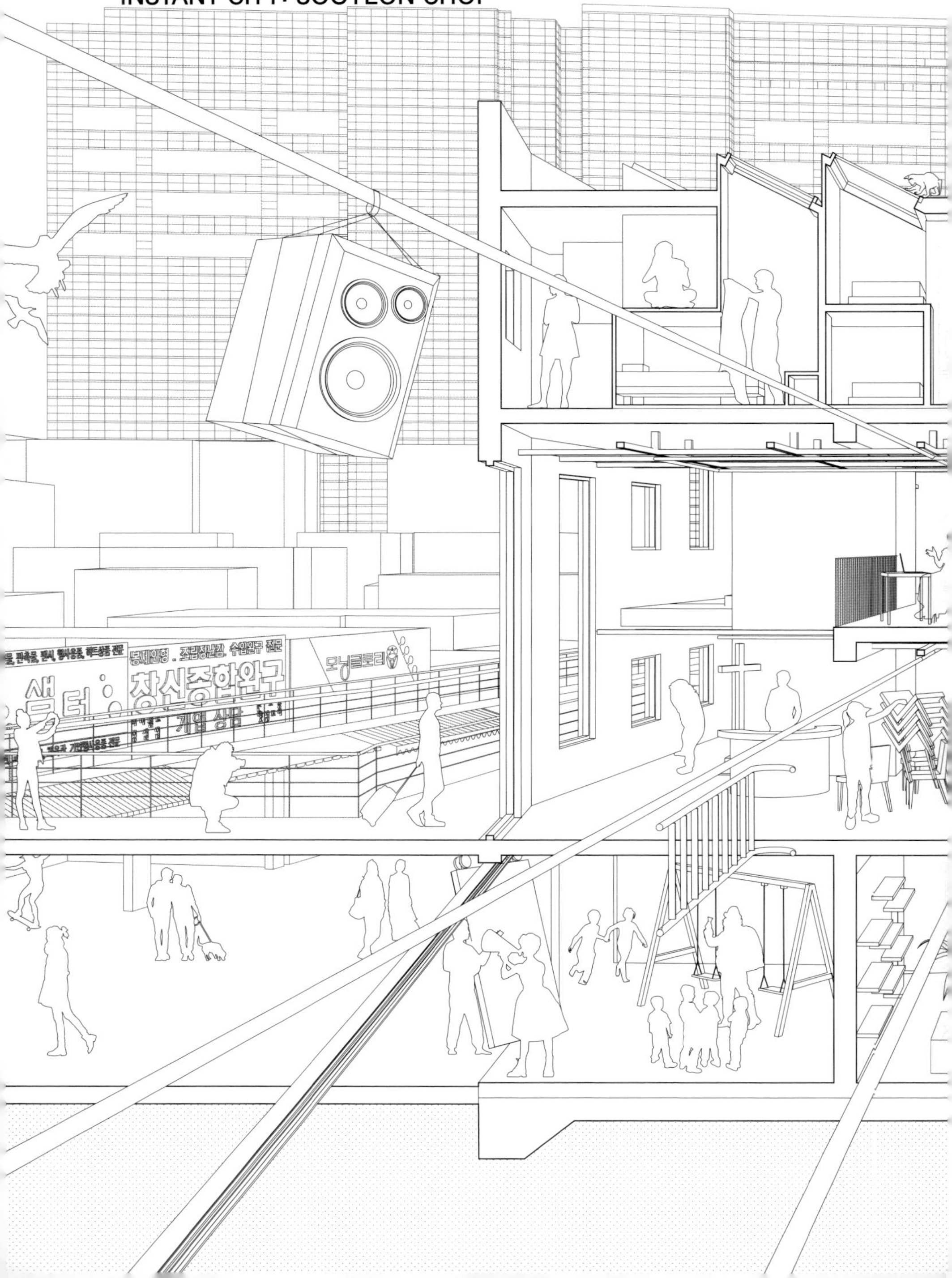

창신동의 일반적인 건물들이 컨테이너처럼 활용되어 프로그램과 용도 사이에 뚜렷한 구분을 두지 않고 현장에서 요구되는 일시적 활동들 사이에 스며들도록 한다. 대상지 전반에 걸친 컨베이어 시스템을 통해 마이크로 주거 유닛과 팝업 임시 매장들로 가구, 상품 그리고 서비스가 모두 배송되며, 이에 따라 건물들 각각이 복합적이고 생동감 넘치는 하나의 도시로서 작동한다.

Instead of making a fixed distinction between program and use, the generic buildings on the site are used as containers that can be infiltrated with temporary activities as needed. Through a site-wide conveyor system, furniture, goods, and services are delivered to both micro-housing units and pop-up businesses so that the buildings themselves operate as a complex and vital city.

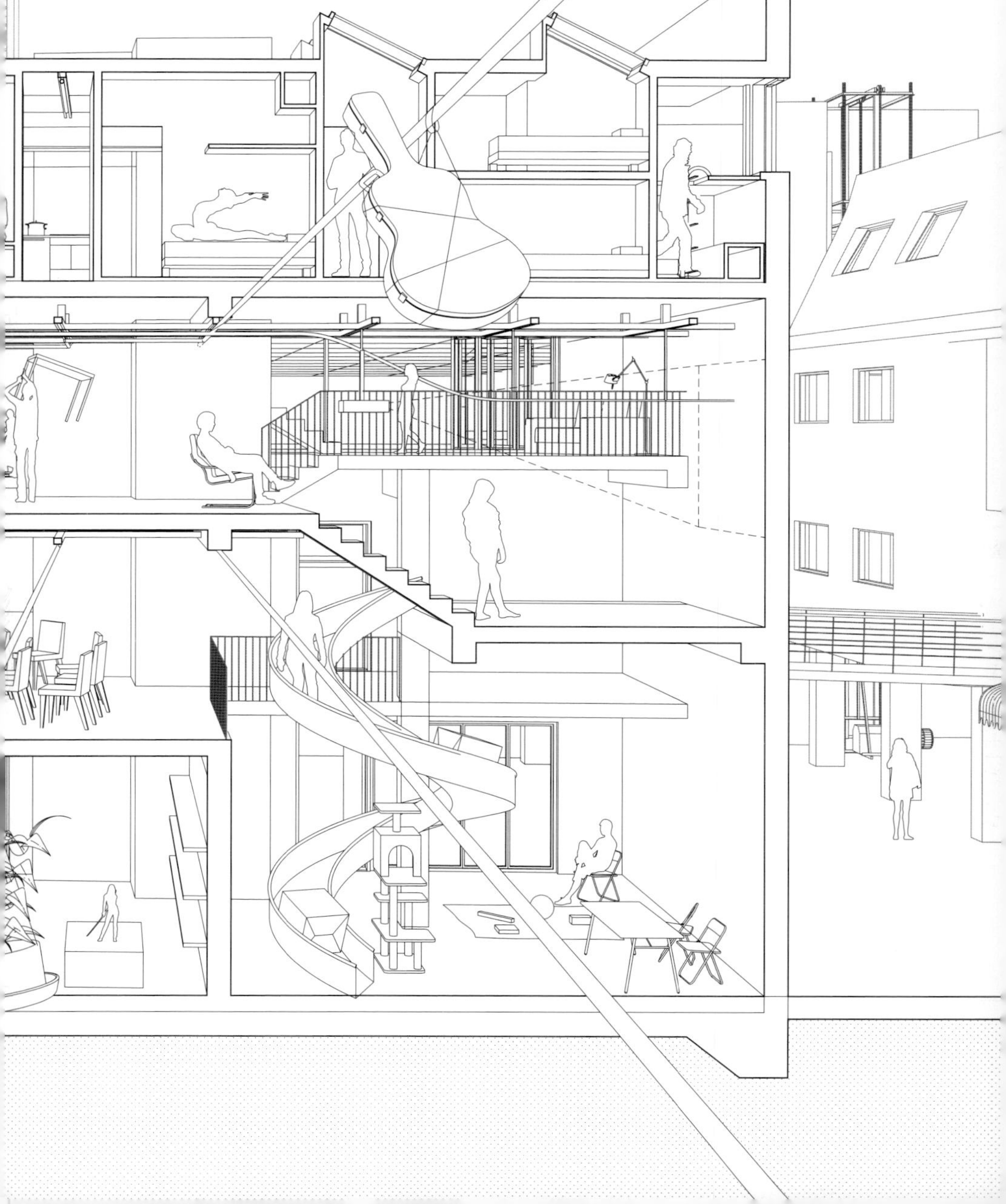

어린이 교육 센터
CHILD EDUCATION CENTER: GRACE MOON

창신동의 규모가 주는 인상은 마치 보행자들에게 좋은 환경일 것처럼 느껴지지만 실제로 현재 거리 환경 속에서 일어나는 활동들은 어린아이들이 안전하게 활동하기에는 어려운 실정이다. 이 프로젝트는 기존의 교통 체계를 다른 곳으로 옮기고 상업 활동이 벌어지는 거리의 바로 뒤에 보행자 전용 구역을 구성한다. 이 새로운 구역은 몬테소리 학교처럼 도시 자체가 배움을 위한 환경이 되게 한다.

Even as the small scale of the site should be conducive to pedestrian infrastructure, the current activities and street layout create conflicts that make it difficult for children to safely inhabit the site. This project diverts existing traffic to create a pedestrian-only zone directly behind the commercial streets. The new precinct works as a Montessori school where the city itself becomes the learning environment.

깊숙한 표면

DEEP SURFACE: DONGHWAN KIM

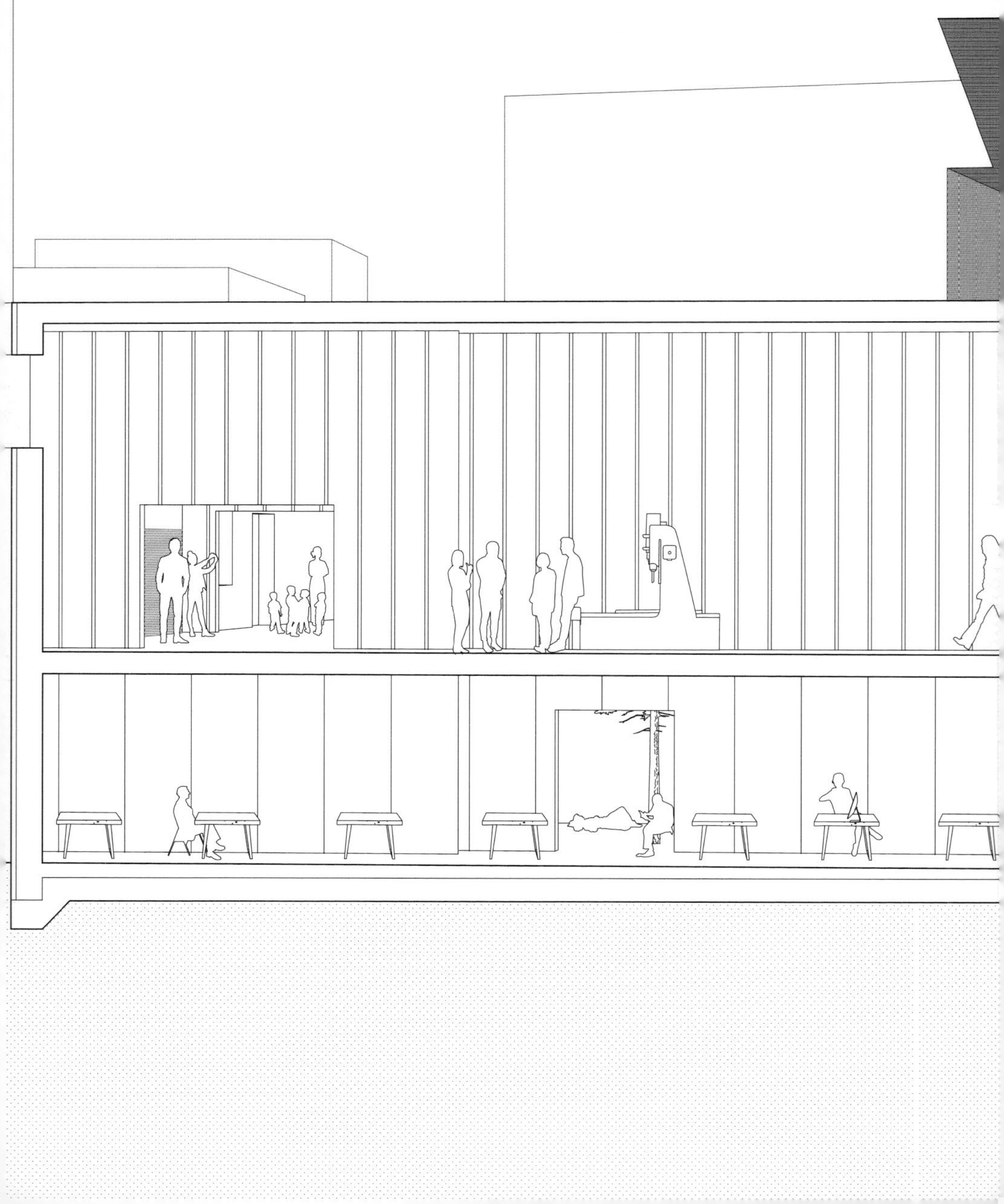

조각처럼 흩어진 기존 건물들을 새로운 방식으로 연결하기 위해 긴밀하게 얽혀진 새로운 기반 시설은 타공 철판으로 된 표면으로 구성된다. 차양, 계단실, 식재를 위한 구조물, 임시 창고 등을 포함한 다양한 기능을 가진 표면들은 시각 기호로 작동함으로써 각 사업체가 대규모의 협력 기관으로 연합할 수 있도록 한다.

To create new connections between the fragmented existing buildings, a perforated metal surface becomes a new intertwining infrastructure. By working in multi-functional ways including shading, staircases, planting structures, temporary storage, as well as performing as a visual symbol, it allows businesses to combine into larger, collaborative organizations.

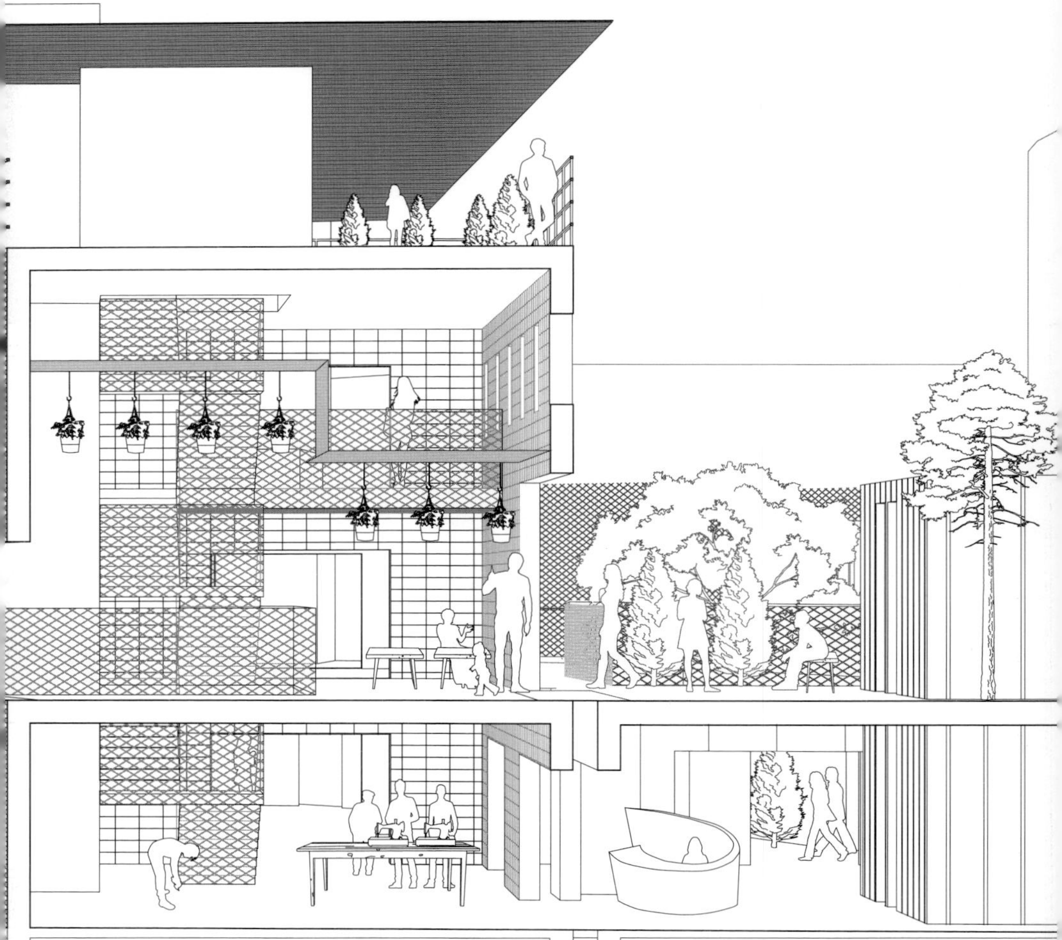

한데 드러난 시간
COINCIDED: SOOHWA HONG

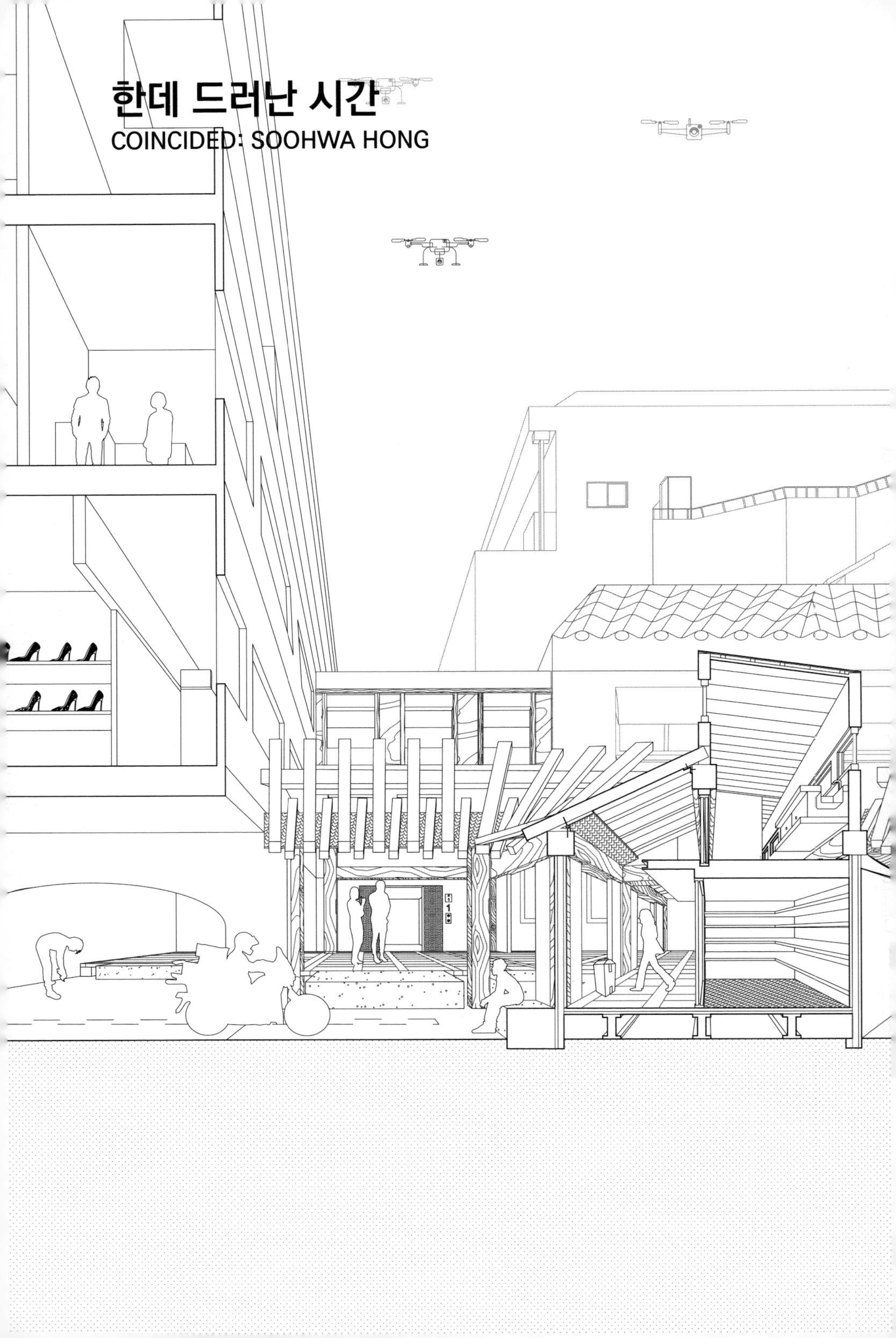

창신동의 현재 용도와 흐름을 분석해보면 한옥을 포함한 70년이 넘은 건물들, 저층의 벽돌 구조물들 그리고 거대 고층 구조인 평화 시장이 주목된다. 저장 공간, 생산 시설, 교육 시설 그리고 휴게 시설을 전략적으로 연결하여 창신동을 연대기적으로 작동하게 한다. 프로그램이 기존의 공간에 걸맞게 하기 위해, 새로운 켜들을 건물들 위로 덮어씌우기보다는 오히려 이전의 켜들을 벗겨 드러나게 한다.

An analysis of existing uses and flows reveals existing buildings spanning 70 years including hanok, low-rise brick structures, and the Pyeonghwa Market megastructure. Through strategic links, warehousing, production, learning, and resting amplify the chronological use of the site. Rather than cover over the buildings, layers are stripped away to conceptually match program to existential space.

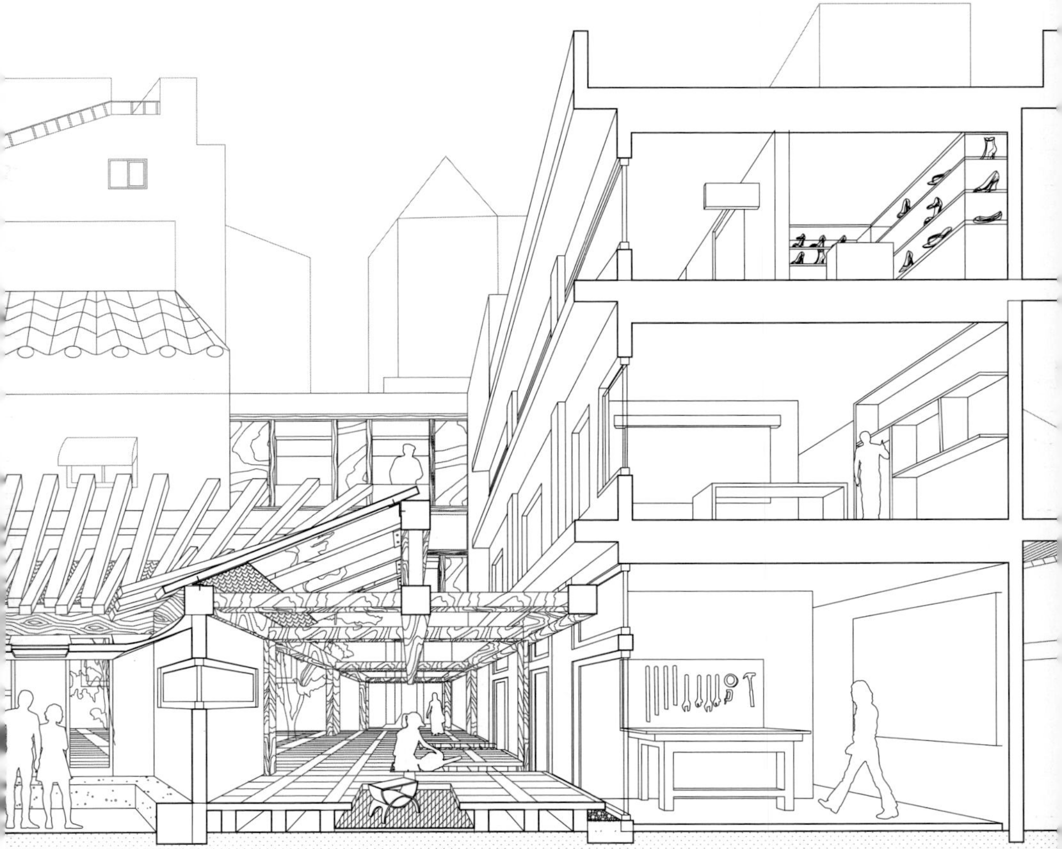

연결된 플랫폼
LINKED PLATFORM: JAESUNG PARK

매우 높은 다양성을 지닌 종로 54길은 제화, 인쇄, 수족관 그리고 문구 등의 다소 관련성이 적은 프로그램들을 포함하지만, 도시 순환 체계에는 중요한 통로이기도 하다. 노반과 지붕 위 높이를 지나는 상징적인 아치형 플랫폼은 실내 설계 디자이너들을 위한 새로운 싱크탱크로서 작동하여, 창신동에서 일어나는 활동들이 조화를 이룰 수 있도록 돕는다.

The extreme diversity on Jongro 54-gil road includes somewhat unrelated activities such as shoe-making, printing, aquariums, and stationery but is also an important circulatory passageway. By creating platforms in subgrade and rooftop levels that are structured by a symbolic archway, the activities in the site are coordinated into a new think-tank for interior designers.

제목 없음

WORKING TITLE: KIERAN MAGUIRE

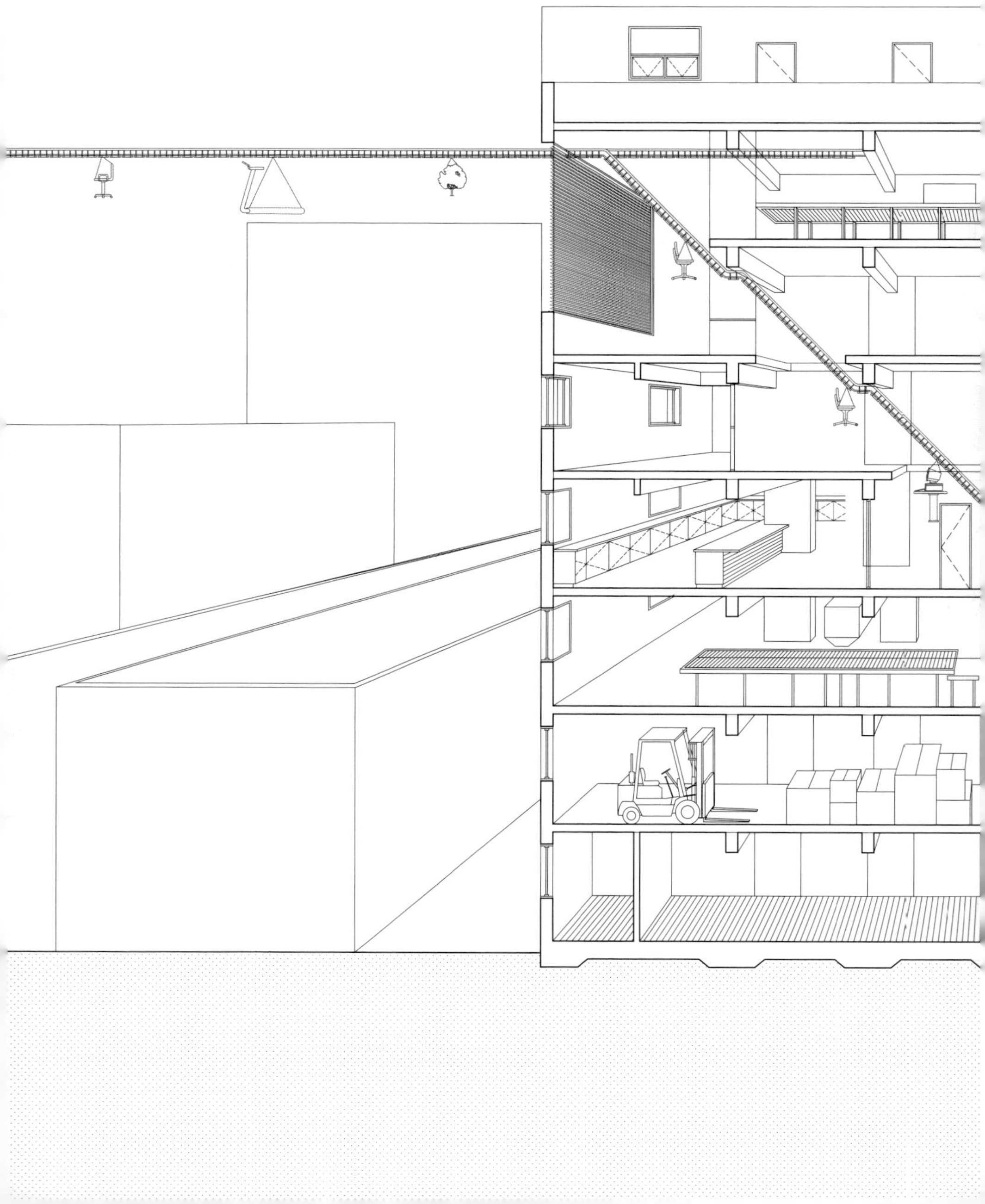

고정된 프로그램을 지정하는 대신 동대문 문구완구시장의 가장 높은 기존 건물 중 하나를 골라 새로운 유통 센터를 설치하기 위해 '제목 없음'이라는 개념을 제안한다. 건물의 전면은 주요 도로와 접해있고 후면은 작은 골목들과 만나는 이상적인 위치에 있다. 창고와 제조 시설을 조직하는 이 새로운 프로그램은 유통 센터와 더불어 도시적 규모의 컨베이어 시스템을 통해 인근의 작은 사업체들을 보조한다.

Rather than propose a fixed program, the idea of 'working title' positions a new distribution center into one of the highest existing buildings in the DDMSTM. Ideally located, the building fronts major streets and backs onto small alleyways. New programs coordinate storage and manufacturing, with a distribution center that serves smaller nearby surrounding businesses through an urban scale conveyor system.

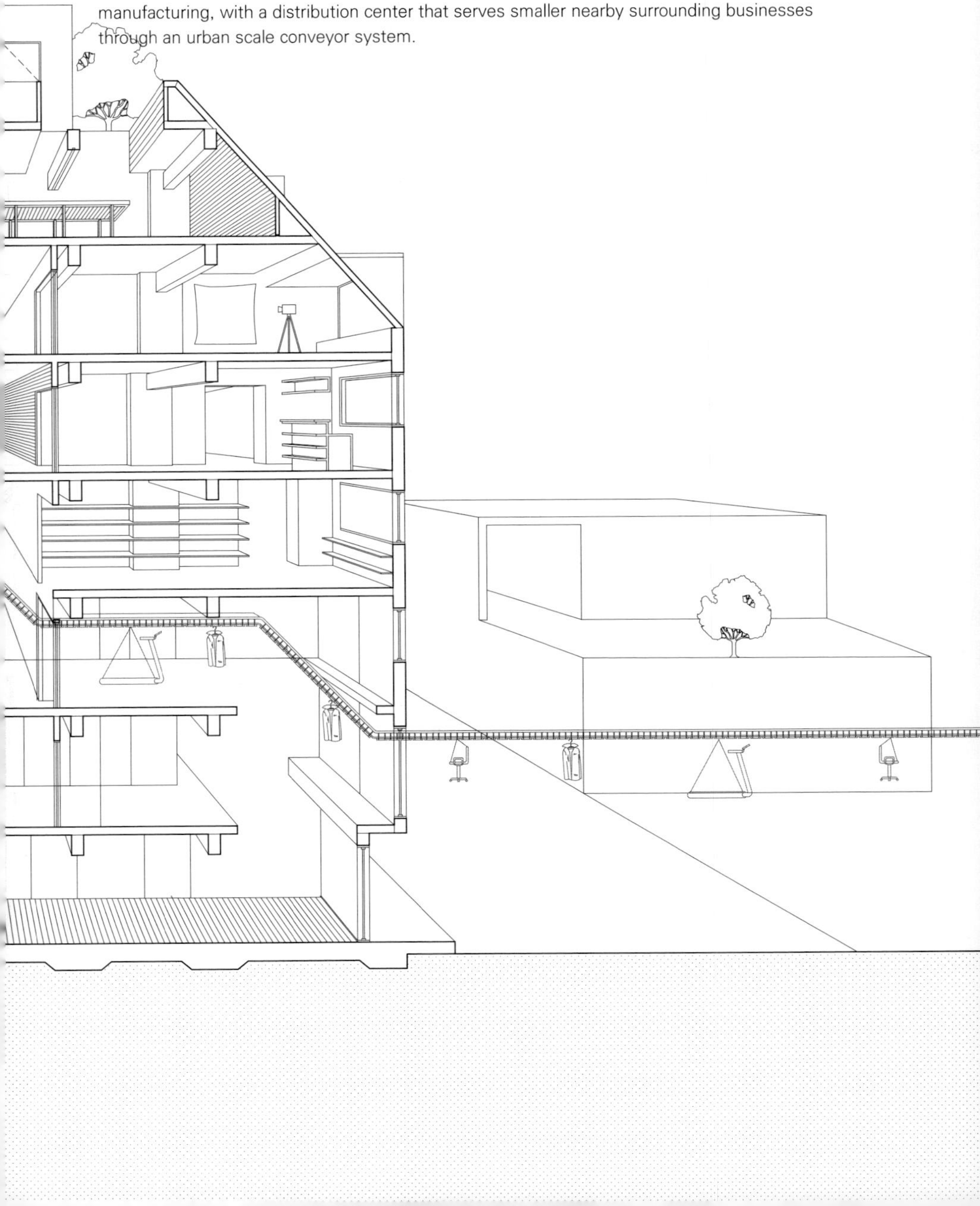

씨실날실 클러스터
INTERWEAVING CLUSTER: DONGHYUN KIM

창신동의 역사적 한옥들은 도시적 맥락과 동떨어진 채 오랫동안 감춰져 있었다. 이 한옥들을 한낱 유적으로 드러내는 대신, 주변의 잠재력을 가진 사업체들과 씨실과 날실처럼 교차로 엮이도록 한다. 다공성을 가진 가벼운 플랫폼이 한옥의 위로 뻗어 나가며 한옥의 고유한 구조 그리고 마당과 관계를 맺어 더 큰 규모의 협력 공간에 대한 필요를 충족시킨다.

The site contains many historic hanok that have been covered over time making them indistinguishable from their context. Instead of merely revealing them as relics however, the project interweaves them with the potential surrounding businesses by spanning a light and porous platform above them that responds to both the hanok's original structure and courtyards and the needs of larger-scale collaborative spaces.

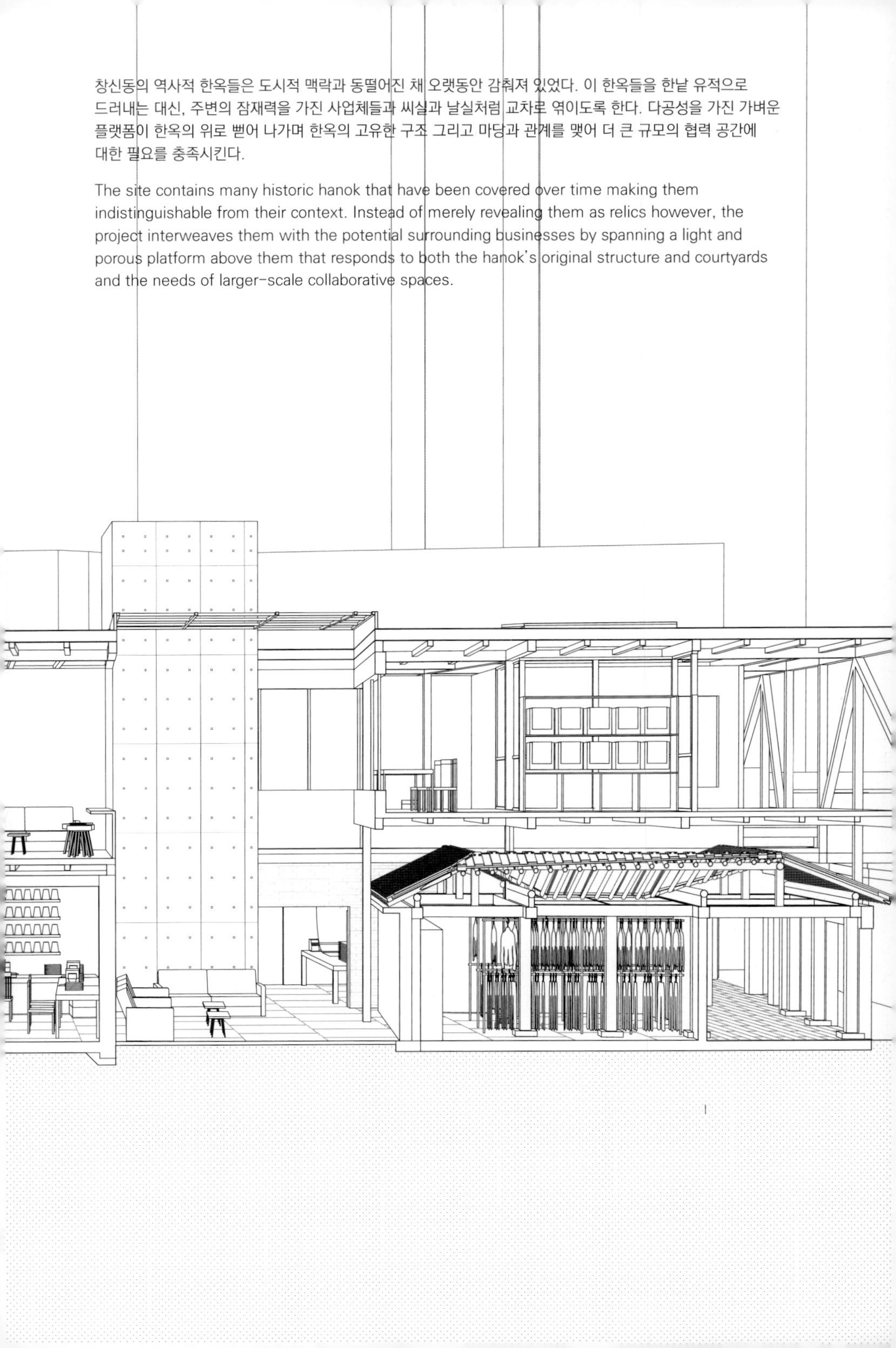

창신 시장 캠퍼스
CHANGSIN MARKET CAMPUS: SANGBEOM KIM

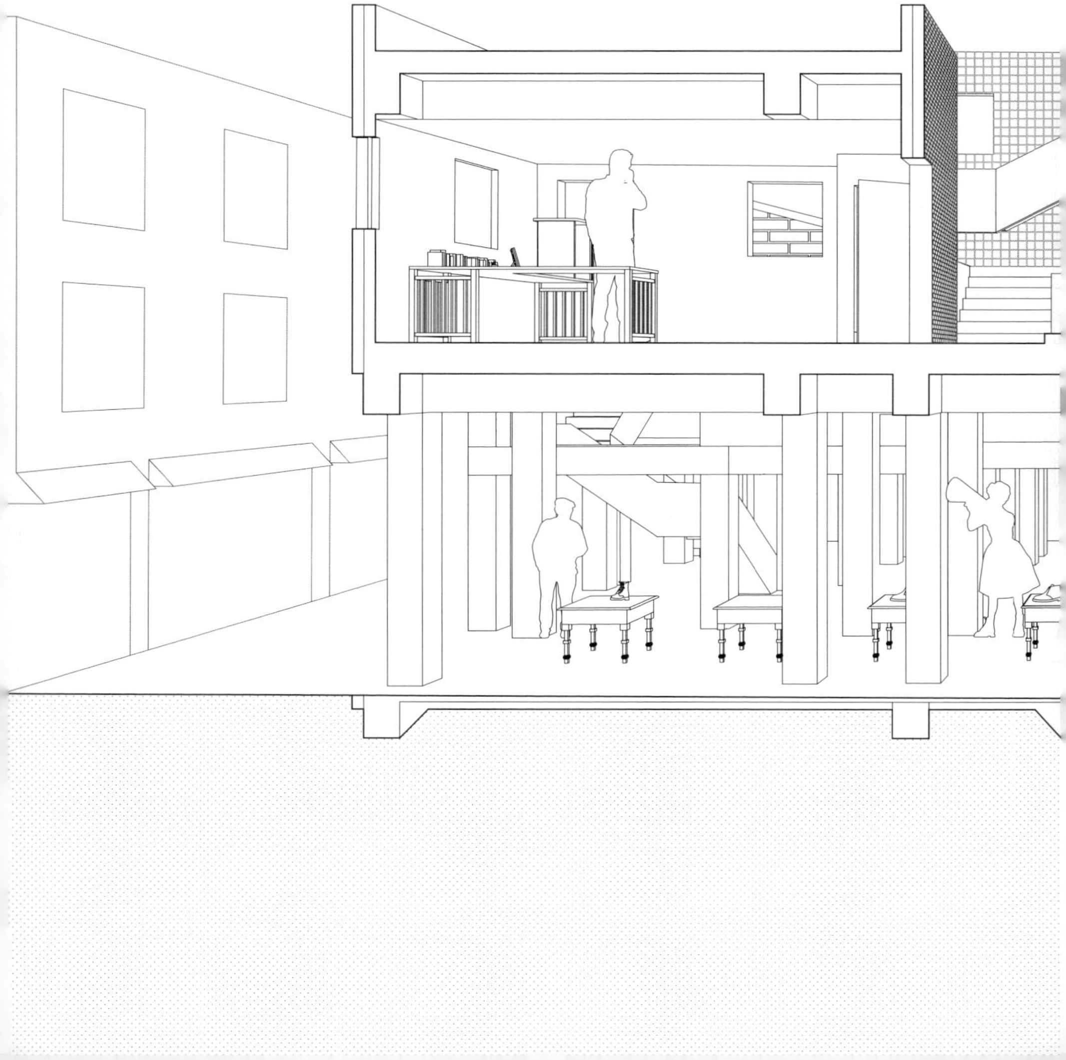

창신동은 높은 밀도 때문에 커뮤니케이션을 조성할 공공 공간이 적고, 따라서 상업적 문화적 거래 또한 제한되어있다. 이 프로젝트는 학생이 관리하고 주도하는 마이크로 캠퍼스를 제안한다. 선택된 건물들의 일층 공간을 다주식 공간으로 활용하여 열린 교환거래 시장을 개최한다. 또한, 현재 비어있는 상층부는 학생이 관리하는 주거시설로 사용하거나 학생 주도의 디자인 워크숍으로 이용된다.

Within the site's density, there is little public space to foster communication, thus limiting both commercial and cultural transactions. This project proposes a new student-managed micro-campus: By creating an open exchange market that combines the ground floor of several buildings into a new 'hypostyle hall' trade area, the upper levels, now currently empty, can host student housing and design workshops.

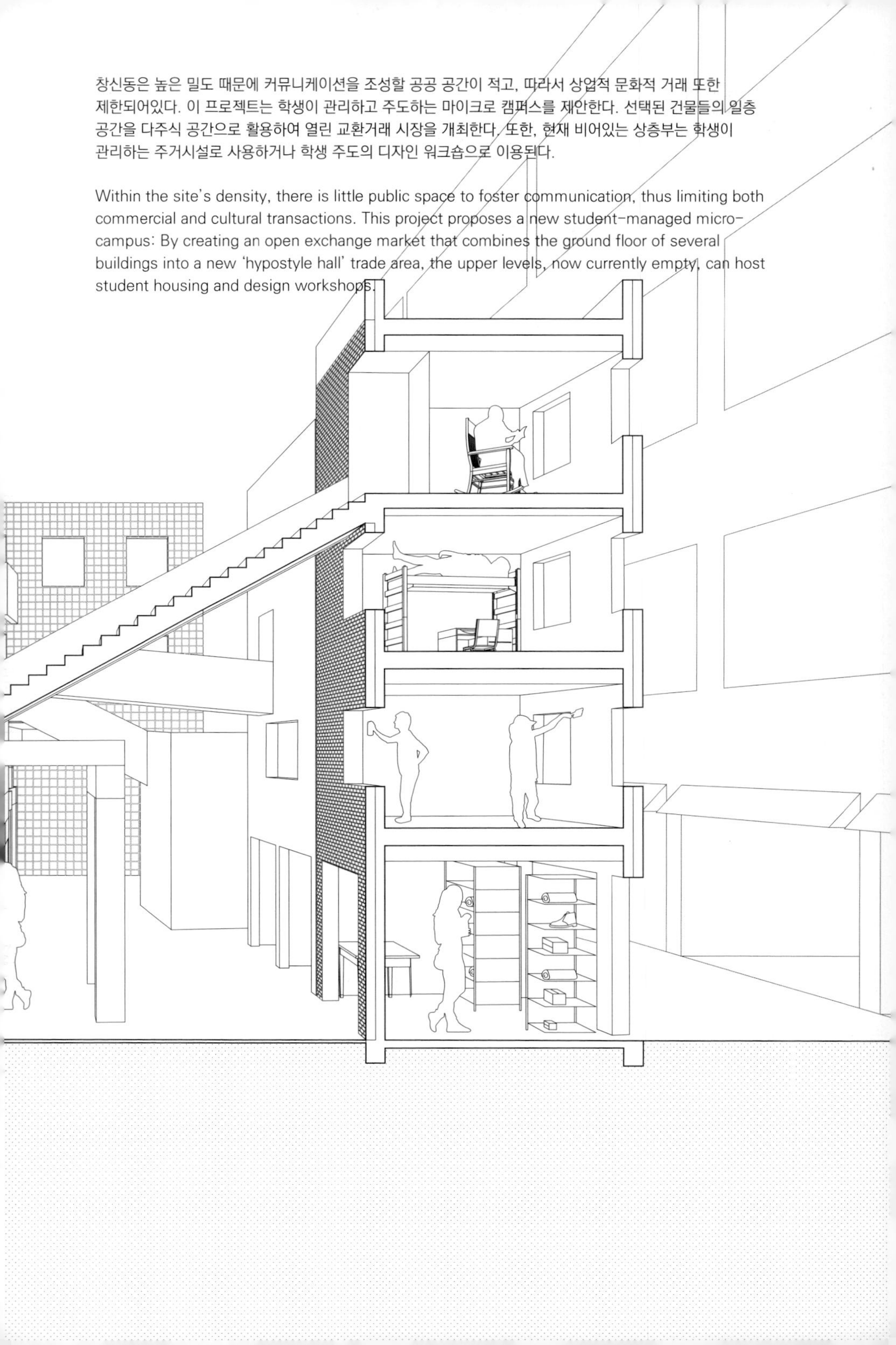

지면 곱하기 지면
MULTIPLYING GROUND: CHANGYONG KIM

대상지 남쪽 면에 선형으로 자리한 평화 시장의 1층과 그 주변 구역을 차지하고 있는 기존의 신발도매상가는 천천히 쇠퇴하고 있다. 2층 이상 높이에서 거주 공간과 상업 공간을 연결하고, 지상에 공공을 위한 보행 도로를 조성하고, 인접한 청계천으로 이어지는 지하 연결로를 설치함으로써, 공간과 프로그램이 밀도 있게 겹쳐져 새로운 문화적 경제적 시너지를 낸다.

The existing shoe market area occupying the ground floor and surrounding areas of the linear Pyeonghwa Market at the southern edge of the site is in slow decline. By creating upper-level connections between residence and commercial, over-ground public pedestrian ways, and sunken connections to the adjacent Cheonggycheon stream, a dense overlay of spaces and programs offers new cultural and economic synergies.

의류 디자인 클러스터
CLOTHING DESIGN CLUSTER: SANGJUN PARK

새롭게 설계한 구조물을 신중하게 삽입함으로서 전체 지역은 하나의 패션 디자인 센터로서 집합을 이룬다. 거주자들은 사적 영역을 유지하면서도 반 층 높이의 공간에서 다양한 프로그램들을 끊임없이 마주치게 된다. 이로써 의류 사업체들은 혁신적인 전매특허 상품을 개발하는 일에 경쟁력을 얻게 된다. 그 과정에서 창신동이 가진 높은 도시적 밀도는 의류 사업체들이 자원을 공유할 수 있게 돕는다.

Through the careful insertion of new structures, an entire block is clustered into a fashion design center. The use of half-levels allows inhabitants to gain teasing glimpses to the diverse programs while still maintaining privacy. This allows design shops to be competitive in developing proprietary and innovative products while the density allows them to share resources.

어울리는 동네
LOCAL INTEGRATION: KYEONGTAI HAN

창신동의 한옥들은 대부분 생활환경의 변화에 매번 적응하기는 어렵기에, 다소 긍정적이지 못한 방식으로 개선됐다. 이 지역에서 부쩍 필요성이 대두하고 있는 학교 그리고 다양해진 가정을 위한 거주 단위들을 삽입함으로써 기존의 유형학적인 마당과 골목들은 새로운 프로그램과 통합되고 재구성된다. 이를 통해 궁극적으로는 보다 역사적인 개념의 보존이 가능하게 한다.

The many existing hanok in the area have not been able to keep up with changes in lifestyle and therefore have been renovated in less positive ways. By injecting a much-needed school and diverse family units into the area, the typological courtyards and alleyways organize the way new unit types and programs are aggregated while allowing for a more historically significant version of preservation.

문지방 공간 다시 쓰기

APPROPRIATION THROUGH TRANSLATION: CAMILA FORERO

대상지의 상업활동에 바로 집중하기보다는 주거 지역, 공동체 그리고 상업 지역 사이의 문지방 공간들에서 공유되는 프로그램을 통해 소통의 수준을 높인다. 사이 공간을 재정의함으로써 현재의 열악한 주택을 철거하거나 장기세입자를 전치 시키지 않고 새로운 편의 시설을 제공한다.

Rather than focus directly on the commercial activities of the site, the liminal spaces between residential, community, and commercial areas become places of heightened communication through shared programs. This redefinition of the in-between brings new amenities to the currently substandard existing residences without having to demolish them and relocate long-term tenants.

새로 사는 동대문

지은이
존 홍

사진·만화·그래픽 디자인
김혜인

번역
강승재
김혜인

지도·단면 디자인
서울대학교 건축설계스튜디오 4-1 (2018)
카밀라 포레로
김창용
김동환
김동현
문서령
박재성
키에란 맥과이어
한경태
김상범
박상준
홍수화
최수연

강승재 (TA)

원제
새로운 주거의 유형들:
서울 구도심의 도시 복합 활용

이 연구는 서울대학교 신임교수 연구정착금으로
지원되는 연구비에 의하여 수행되었음

ISBN:978-89-5592-411-4

NEW FORMS OF LIVING

AUTHOR
John Hong

PHOTOS·CARTOON·GRAPHIC DESIGN
Hyein Kim

TRANSLATION
Seungjae Kang
Hyein Kim

MAPPING·SECTION
SNU DAAE STUDIO 4-1 (2018)
Camilia Forero
Changyong Kim
Donghwan Kim
Donghyun Kim
Grace Moon
Jaesung Park
Kieran Maguire
Kyeongtai Han
Sangbeom Kim
Sangjun Park
Soohwa Hong
Sooyeon Choi

Seungjae Kang(TA)

ORIGINAL TITLE
New Forms of Living: Mixed-Use
Urbanism for Seoul's Historical Center

This work was supported by Research
Resettlement Fund for the new faculty of
Seoul National University

ISBN:978-89-5592-411-4

발행인
김기현
발행처
시공문화사 (*Spacetime*)
출판등록
1993년 3월 12일
주소
서울시 서대문구 독립문공원길 13 (03733) 극동프라자 5층
전화
02)3147-1212
팩스
02)3147-2626

http://www.spacetime.co.kr
spacetime@korea.com